SECRETS OF
BIRD LI

CW00373388

# SECRETS OF BIRD LIFE

*Ron Freethy*

A GUIDE TO BIRD BIOLOGY

BLANDFORD

**Blandford**
An imprint of Cassell
Artillery House, Artillery Row, London SW1P 1RT



First published 1982 as *How Birds Work*
Revised edition published 1990

Distributed in the United States by
Sterling Publishing Co. Inc.
387 Park Avenue South, New York, NY 10016-8810

Distributed in Australia by
Capricorn Link (Australia) Pty Ltd
PO Box 665, Lane Grove, NSW 2066

**British Library Cataloguing in Publication Data**
Freethy, Ron
[How birds work.] Secrets of bird life.
1. Birds. Physiology
I. [How birds work.] II. Title
598.2′1

ISBN 0-7137-2154-5

Typeset by Asco Trade Typesetting Ltd., Hong Kong

Printed in Hong Kong by South China Printing Co. (1988) Ltd.

# Contents

# Acknowledgements

During the production of this book I have been ever-conscious of the debt I owe to a number of people who have given cheerful help and constructive criticism.

Dr. Jim Flegg kindly read the manuscript and such is the depth of his ornithological knowledge that I was pleased to make several alterations at his suggestion. To Beth Young and the staff at Blandford Press I record my thanks for their patience and good humour. I am particularly grateful to Ray Hutchins who undertook the artwork at short notice and I am delighted with the results. To the various photographers I also extend my thanks.

A number of friends kindly read the manuscript including Pat Richardson, Joan Morrisey and Brian Lee. To my son Paul I express my thanks for his eagle sharp eye in correcting proofs. To my wife Marlene who typed the manuscript from my handwriting I express not only thanks but sympathy. She also made several valuable suggestions during the writing of the manuscript. The end product is therefore the work of many people, but the errors which do remain are my own responsibility which I gladly accept.

# List of Tables

# List of Illustrations

# Introduction

This book is intended as an introduction to the science of ornithology. I have assumed no deep biological knowledge in my readers, and have kept technical terminology to a minimum. There was once a preacher whose congregation complained at the length of his sermons. 'Brethren,' he replied, 'if thou knewest how much I left out thou woudst not complain at what I include.' I felt in much the same position when compiling the first two chapters, those on evolution and classification. In the first, with many well-documented fossil birds to choose from, my selection had to be short and was to some extent subjective. In Chapter 2 also I have, for example, described the very important passerine order in much the same space as that devoted to other much smaller orders. I have compensated for this, at least in some degree, by providing an extensive bibliography. In a book of this nature there are bound to be points of contention especially with regard to precise classification. Taxonomy is always changing and is always a thorny area of debate, and I have in many cases presented my own opinion on certain issues. Where, however, there are significant differences of opinion, as in the placement of the tinamous and the hummingbirds, for example, I have tried to present a balanced overview.

For the rest of the book I have followed what seems to me a logical order of presentation, my reasoning being as follows: Chapter 3 deals with the general anatomy of the bird including details of feather structure, thus leading into the fourth chapter which concentrates on flight. In order to achieve efficiency as a flying machine the avian body needs liberal supplies of both oxygen and food, and these topics are the subjects of my fifth and sixth chapters. To control these complex processes a highly evolved nervous system is required: this is dealt with in Chapter 7. This nervous system also controls the breeding cycle (Chapter 8) and migration (Chapter 9) and other aspects of behaviour (Chapter 10). In Chapter 11 we return to a wider perspective on the bird world, to look at the zoogeographical zones and the distribution of birds, and with this in mind we can proceed to Chapter 12, a discussion of man's relationship with the birds of the world.

At the conclusion of the book I hope that the reader will be able to understand what makes a bird 'tick' and that his pleasure in ornithology may be thus enhanced.

*Chapter One*

# The Evolution of Birds

## What is a Bird?

In one sense it is in an effort to answer this question that this book has been written. A bird is technically defined as a warm-blooded backboned animal having two pairs of limbs based on a pattern of five 'fingers', the front pair of limbs having been modified into wings. The skin is covered by feathers except on the legs where there are horny scales, a relic of reptilian ancestry.

## How did Birds Evolve?

During the course of evolution the techniques involved in powered flight have been mastered four times: by insects, reptiles, birds and mammals. The birds were not the first nor were they the last creatures to come up with the necessary solutions.

Insects were the first into the air and during the Carboniferous period (*see* Table 1) they played the same role in the complex network of life now assumed by birds. The usual term for the position occupied by a particular species within its environment is its ecological niche.

The second group to solve the problems of flight were the reptiles which have been appropriately named pterosaurs ('flying reptiles') but they have also been called pterodactyls ('flying fingers'). The earliest fossils so far discovered date from the latter part of what is known as the Jurassic period—some 170 million years ago. The pterodactyls have the distinction of having produced the largest known flying animal: it was called Pteranodon and had a wingspan of over 8 m (27 ft). By the late Cretaceous period, some 80 million years ago, however, the flying reptile experiment had failed. There was a period of almost 100 million years when the third group of animals to evolve powered flight—the birds—overlapped the pterodactyls, and the two groups were in competition for the same ecological niche. The first birds were in existence from about 150 million years ago, during a period referred to as the Upper Jurassic.

There are many heated disagreements regarding the precise lines of bird evolution, but all workers seem agreed that birds evolved from reptiles and have come from the same stock as the dinosaurs: the arguments centre around the route which the development followed. One suggestion is that the ancestor was arboreal and merely glided from one tree to another, and if it landed on the ground it literally clawed its way up a tree in preparation for take-off. Another view is that the ancestor was a ground-dwelling, jumping animal which extended the length of its jumps by the gradual development

Table 1
GEOLOGICAL TIME CHART

| | | Years distant (m = million) |
|---|---|---|
| CENOZOIC ERA | | |
| Quaternary Period | Holocene | 1 m |
| | Pleistocene | 2 m |
| Tertiary Period | Pliocene | 7 m |
| | Miocene | 26 m |
| | Oligocene | 38 m |
| | Eocene | 54 m |
| | Palaeocene | 70 m |
| MESOZOIC ERA | | |
| Cretaceous Period | | 135 m |
| Upper Jurassic Period | | 150 m |
| Lower Jurassic Period | | 195 m |
| Triassic Period | | 225 m |
| PALAEOZOIC ERA | | |
| Permian Period | | 280 m |
| Carboniferous Period | | 345 m |
| Devonian Period | | 410 m |
| Silurian Period | | 440 m |
| Ordovician Period | | 530 m |
| Cambrian Period | | 570 m |
| PRE-CAMBRIAN ERA | | 2800 m |

of elongated forelimbs. So far, so good, but we are still left with two significant developments to consider. Firstly, how did feathers evolve, and secondly, how was the development of warm blood in birds achieved? The first question is much more easily answered than the second. Reptilian scales appear to have gradually elongated and flattened to support the weight of a flying creature, eventually becoming finely divided (feathery in fact) to enable them to absorb high pressures without permanent damage. With regard to the evolution of a warm-blooded bird from a cold-blooded dinosaur we find many more problems, although recent evidence would seem to suggest that we may have made a bit of a mountain out of a scientific molehill. Let us look at the difference between warm- and cold-blooded animals: a 'cold-blooded' animal (called poikilothermic) has little, if any, control over its body temperature. This means that when the environmental temperature is high the animal's blood temperature can rise quite dramatically, and when the environmental temperature falls then the metabolism can become so slow that the animal may not be able to survive periods of prolonged cold weather unless it hibernates (or migrates over huge distances). In contrast, warm-blooded animals (called homoiothermic), the birds and mammals, are able to maintain a constant body temperature, whatever the environmental conditions encountered: this obviously gives a great deal of independence of movement and is of great survival value.

Let us now return to pre-avis, ancestor of our feathered tribes. How did it evolve from a cold-blooded reptile? The answer may be that it did not. Alan Charig has suggested that some dinosaurs may have been warm-blooded, and could therefore have evolved into birds much more logically than has often been imagined. There will always be conflict of opinion, but the fact that dinosaurs may have been warm-blooded and are the ancestors of the first bird, could well lead to the speculation that dinosaurs did not all become extinct—they merely evolved into birds and are with us still. 'Amateur' palaeontologists always seem to demand a missing link, and to some degree the reptile/bird frontier does have its own check-point-Charlie. This is Archaeopteryx, the name translating as 'ancient wing', and it is the oldest bird fossil so far discovered.

## ARCHAEOPTERYX

The story of the discovery of Archaeopteryx takes us to Bavaria in southern Germany. Here huge deposits of limestone were laid down some 150 million years ago; climatic conditions during Jurassic times seem to have been warm and humid with huge lagoons and with lush vegetation covering even the firmer areas. The limestone is of particularly fine quality and from about AD 1800 was in great demand for lithographic work. This process demands that the working material must be free from any sort of imperfection and it is therefore split into slabs, carefully and skilfully inspected, and any fossils removed along with any other offending object. The area became famous both for the quality and quantity of its fossils.

In the year 1861 a fossil was discovered and passed to one Herman von Meyer who examined it closely enough to discover the imprint of an avian feather. After an excited search an imperfect, but obviously feathered skeleton was found and named *Archaeopteryx lithographica* (*see* Fig. 1a). Here, then, was the first feathered bird. The news was not slow to travel. It passed from von Meyer to a keen and able palaeontologist named Haberlein, who was a local doctor, and often treated the quarrymen who would pay him in fossils rather than in cash. In 1862 the British Museum obtained the specimen and still have it, but keep it in great secrecy. The specimen actually on display in the Natural History museum is a copy made out of glass fibre.

The world's 'bird scientists' now have four and a bit fossils on which to base their theories. A second, and very splendid, specimen was unearthed in 1877 and is now in Berlin. It was not until 1958 that a further specimen was found within a slab of rock leaning against a workman's hut, and this is now kept in a Bavarian museum. In 1970 another skeleton was found by a Dutch

**Fig 1a** Archaeopteryx fossil

scientist whilst working on collections of pterodactyls. Finally, what of the original feather which consisted of a part and counterpart? One half is housed in Munich and its 'mirror image' is kept in Berlin.

Let us now consider what these fossils have revealed. There is no doubt whatever that Archaeopteryx was a bird, but no doubt either that it was very primitive, still retained many reptilian features, and had not evolved many features typical of modern birds (*see* Table 2).

Despite the bulk of the evidence pointing to Archaeopteryx the bird, we can still trace a resemblance to the dinosaurs, but from which branch has it

**Fig 1b** Archaeopteryx reconstruction

ascended? It has been discovered that very few dinosaurs have collar bones which, as shown in Table 2, are fused together as the wish-bone; this is found both in Archaeopteryx and modern birds. The group of dinosaurs with collar bones are known as coelurosaurian dinosaurs and it would seem that it was from this line that Archaeopteryx evolved. Furthermore, a comparison between the forelimb skeleton of Archaeopteryx and that of a coelurosaurian are very similar. Indeed it would seem that the first known bird was only a weak flier and may even have had to climb trees in order to get airborne. Yet another similarity can be found on the examination of the feet. Some aspects

Table 2
SHOWING THE CHARACTERISTICS OF ARCHAEOPTERYX COMPARED TO BIRDS AND REPTILES

| Non-Bird like (Reptilian) Characteristics of Archaeopteryx | Bird-like Characteristics Possessed by Archaeopteryx | Bird-like characteristics not possessed by Archaeopteryx |
|---|---|---|
| Presence of a long, bony tail. (This contained separate vertebrae and was not due to long tail feathers.) | Presence of feathers (not found in any other class except birds). Even primary and secondary feathers can be detected. | The presence of a keel to which huge flight muscles are attached. |
| The presence of teeth. | Presence of a wish-bone formed by fusion of two collar bones (unique to birds). | The presence of air spaces in the bones to lighten the body weight and to provide reserve oxygen supply. |
| The presence of claws on the ends of fingers. | Presence of wings. | Presence of a large brain to co-ordinate long, active flights. |

of the coelurosaurian 'running foot' have been retained by Archaeopteryx, and the idea of a running jump into the air by our bird/dinosaur cannot be ruled out.

FOSSIL BIRDS SINCE ARCHAEOPTERYX

Bird palaeontology has not made such dramatic progress as has been the case with mammals. Man obviously has a vested interest in following the evolution of mammals, and therefore has tended to regard bird fossils as something of an evolutionary backwater. Secondly, and much more importantly, bird bones are hollow, can therefore be crushed and do not fossilise easily. This explains the huge time interval which elapsed between the Jurassic Archaeopteryx and the birds so far identified from the Cretaceous period which spans from 135 million to 70 million years ago.

BIRDS OF THE CRETACEOUS PERIOD

An important ecological constrast between birds of this period and Archaeopteryx is that by far the majority of Cretaceous remains were found in the marine environment and in the chalk deposits resulting from this. Up to 1978 some 23 fossil birds had been identified from the Upper Cretaceous, and included in the list are Ichthyornis (the 'fish bird') and its contemporary, Hesperornis (the 'dawn bird').

*Ichthyornis*

Reconstructions of this species (*see* Fig. 2) reveal it to be very tern-like and some 20 cm (8 in) in length. Despite its great antiquity there are some

**Fig 2** Ichthyornis **Fig 3** Hesperornis

surprisingly modern features in its make-up; not least amongst these is the large prominent keel to which powerful flight muscles would have been attached. Here then was a much more efficient flying machine than Archaeopteryx.

The name 'fish bird' would seem to suggest that the primitive reptilian feature of having teeth had been retained, the eating of fish seeming to require this property; but recent work carried out by Dr. J.T. Gregory has resulted in a rethink. He noted that only the lower jaw of Ichthyornis was found to possess teeth, and this jaw resembles that of swimming reptiles called mosasaurs so closely that confusion in identification cannot be ruled out. Attention, not unnaturally, turned to the upper jaw fragments, and these do show structures which have been called alveoli, but they cannot be said to prove the certain existence of teeth. Thus the fish bird may well turn out to be toothless *but* what can be said with certainty is that Ichthyornis was an efficient flying machine.

*Hesperornis* (*see* Fig. 3)

With regard to the dentition of the 'dawn bird' we can speak with much more certainty. There can be no dispute that this bird possessed teeth, certainly on the upper jaw, and many workers think that the teeth on the lower jaw are authentic although Swinton has voiced the opinion that they may also be associated with mosasaurian reptiles.

When compared with Ichthyornis (*see* Fig. 2), the dawn bird was large indeed and reached a length in the order of 180 cm (6 ft) which fact begs the question—how did it fly? The simple answer is that it did not. This merely underlines that during the Cretaceous period the avifauna was already considerably diversified, which from an evolutionary point of view was a very satisfactory state of affairs. A close examination of Hesperornis reveals that the skeleton is very well developed but the shoulder girdle does seem rather weak, the breast bone lacks a keel and many of the wing (arm) bones are missing. Many workers have suggested that the bird had no wings at all, but there seems no justification for this although the wings were certainly vestigial. This obviously placed an extra burden upon the legs, and these were, as one would expect, very powerful structures; but they seem to be developed so far along the road towards aquatic living that the bird may have been incapable of efficient movement on land. Here then was the almost complete waterfowl. Other birds thriving around this time included Baptornis, an early relative of the grebes, and Caenagnathus which, judging from its skeleton, would not seem to have been a flier or a swimmer, but an out-and-out land runner.

## BIRDS OF THE TERTIARY PERIOD

By the time the end of this period was reached, some three million years ago, many of the families of birds we know today were already in existence. The Tertiary was a period of intense geographical change. Temperatures were high, vegetation lush and the speed of evolution at this time was at a maximum.

Birds had already evolved into the two distinct structural forms which are with us today. These are the fliers, which have been called the carinates, and those which have abandoned flight altogether in order to concentrate their evolutionary energies on becoming fast runners, the ratites.

During the Tertiary period there seem to have been some fearsome birds at large in the countryside. Many old legends relating to such flying slaughter-houses as the predatory roc in *Sinbad the Sailor* may well have had some substance in fact, their bloodthirsty deeds being somehow incorporated into our folk memory. Certainly present were the moas of which one giant called *Dinornis maximus* stood 3.5 m (nearly 12 ft) in height. One bird from this group—the Kiwi—is still in existence. It is a fowl-sized bird with its nostrils situated at the end of its bill and seeming to possess a well-developed sense of smell.

Of the carinates, fossil evidence is available to show that penguins, divers, grebes, rails, albatrosses, pelicans, cormorants, herons, cranes and gulls were already thriving, and from these times onwards it is possible to recognise all the bird orders making up the modern system of classification (*see* Chapter 2). Before turning our attention in this direction we need to discuss the extinction of some species and the evolution of the new ones to take their place. Two species which evolved during Tertiary times have become extinct relatively recently, and have received a great deal of publicity. These

**Fig 4** The great auk or garefowl

are a huge pigeon, the dodo, and a relative of our modern day razorbill, the great auk *Pinguinus impennis*.

*The Great Auk*

It is worth describing this species in some detail because it is the one extinct species of bird which is included in the first 'handbooks' of birds. Here then is a link between the palaeontologist and the modern birdwatcher.

The species was usually referred to as the garefowl (*see* Fig. 4). It was a large bird quite unable to fly, laying a single very nutritious egg, and whose flesh was said to be excellent. Its oil was used as fuel for lamps, and its feathers were another valued source of revenue. There can be little doubt that the bird was at one time hunted by the people of the islands of St. Kilda whose economy was based totally upon seabirds.

The historian Martin Martin makes reference to it and says that the bird 'lays its egg on the bare rock and has a hatching spot on its breast'. Martin's record, written in 1697, is not an eye-witness account, but he had quite obviously spoken to people who had seen the bird alive. Since Martin's day there have been several possible garefowl sightings, the most recent one being in 1840. Two St. Kildans are said to have caught a great auk on Stac an

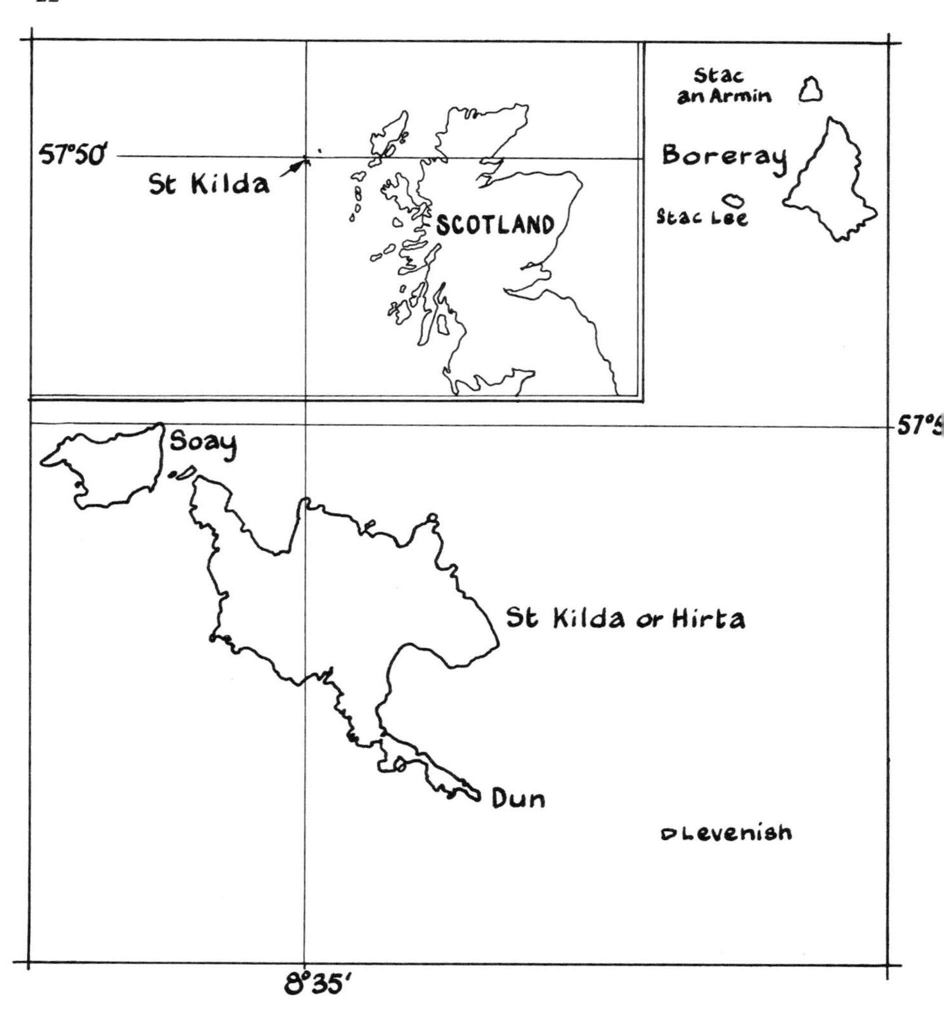


**Fig 5** Map of the St. Kilda group showing the main island, Hirta, and the group's position off the north-west of Scotland

Armin (one of the islands making up the group), but they cannot have been very familiar with the species for they accused it of being a witch directly responsible for a spate of storms which had devastated the islands; found guilty, the bird was stoned to death. This record, perhaps quite rightly, is not accepted by all authorities since the remains were never recovered. The last authenticated record of the garefowl was one killed at Waterford harbour in Ireland in 1834, and this bird now has a distinguished perch in the museum of Trinity College, Dublin.

The last breeding record would seem to be from Orkney in the year 1812. I have an 1840 copy of Maggillivray's *Manual of British Birds*. This was, in its day, equivalent to the popular handbooks which folk purchase today to take away on seaside holidays. It always gives me a peculiar feeling to read about a bird we now know to be extinct, as if we might just be lucky enough to see one on our next birdwatching expedition. I will let Professor Maggillivray describe it to you.

> Length about 30 inches [75 centimetres] wings diminutive with the quills scarcely longer than the coverts; the tail short, of fourteen feathers; bill rather longer than the head, black with eight or nine white grooves on the upper, ten or twelve on the lower mandible; the head neck and upper parts black, the throat and sides of the neck tinged with chocolate brown; the wings with greyish brown, the upper parts glossed with green; the lower parts and a large oblong spot before each eye with the tips of the secondary quills white . . . This species is met with at high latitudes, along the coasts of both continents, but not in great numbers. A few individuals have been seen about the islands of St. Kilda and our North Eastern islands. One was captured in 1822, but made its escape. The habits of this remarkable bird are little known. It is supposed rather than observed to be incapable of flying.

We have learned a little since Maggillivray's day. The white eye-spots, for example, were absent in the winter plumage, and we can assume that they had some part to play in the breeding cycle. The great auk, like most of its tribe, was a colonial breeder, but in winter it is spread far and wide—also typical of auk behaviour. Without doubt the story of this most interesting species is a sad one. It perished as a result of the lethal combination of man's cruelty and greed. If we learn our lesson from past mistakes then the great auk and others like it may not have died in vain. So much for the decline and fall of an ancient species: we must now consider how new species evolve to replace those who fail to adapt to changes in their environment.

## How Species Evolve

There are three main ways in which new species may evolve: by mutations; by natural selection; because of geographical isolation.

### MUTATIONS

A mutation is described as a sudden change in the genetic make-up of an organism. This can arise by chance or be induced by radiation or perhaps even by chemical means. Each species is firmly based upon what is actually a very ancient tried and trusted blueprint and any sudden change will usually be harmful and often fatal. It is more than likely that any mutation will not be passed on to the next generation. Just occasionally, however, the mutation is beneficial, is therefore retained and the species changed.

NATURAL SELECTION

During the course of time each individual has become adapted to its own particular ecological niche. A black-headed gull, for example, has evolved through many generations and can cope with its chosen environment. Thus there inevitably develops an optimum wing length for successful flying in these conditions. An individual whose wing length is far too long or far too short will fail to compete within the species niche, will be less likely to breed and so the species will maintain the status quo. Individuals which differ markedly will almost always perish, but if several manage to survive they may find an unoccupied niche and the formation of a new species will be possible.

GEOGRAPHICAL ISOLATION

This was regarded by Charles Darwin as practically the only way in which new species could evolve, although he did not fail to realise that, once isolated, a population of organisms would be subjected to some degree of natural selection whenever any form of ecological stress built up, be it food shortage, increase in population, predator pressure or whatever. Let us look at Darwin's theory from his observations on the Galapagos Islands and then go on to see how this can account for some of the variations of birds found in Britain and other offshore islands.

*The Galapagos Islands* (*see* Fig. 6)

Situated some 605 miles (970 km) off Ecuador, the Galapagos which straddle the Equator are so isolated that they constitute a perfect evolutionary laboratory. The islands—there are 19 which have been named plus 42 smaller islands and countless even smaller islets and rocks—are, in geological terms, very young. They were thrust up from the depths of the Pacific as a result of volcanic action some one to three million years ago. Galapagos is the Spanish word for a tortoise and it was a race of giant tortoises found on the islands which set Darwin's mind working. He noticed a basic similarity to mainland forms, but there were subtle differences; furthermore there seemed to be no mixing between the tortoises of the various islands; and differences between their individual populations were apparent. Darwin also noted differences in bird life including the presence of flightless cormorants, but it was on the 26 kinds of land bird which he found there that the naturalist was able to exercise his mind concerning the origin of species. What have now become known as Darwin's finches are easily the most common of the land birds occurring in the Galapagos. They are all similarly coloured a dull brown and are separated on the basis of bill shape and feeding habits (*see* Fig. 7).

Firstly we have the six species of ground finch, which occur mainly on the lowlands and coastal zones; as their name implies they are predominantly

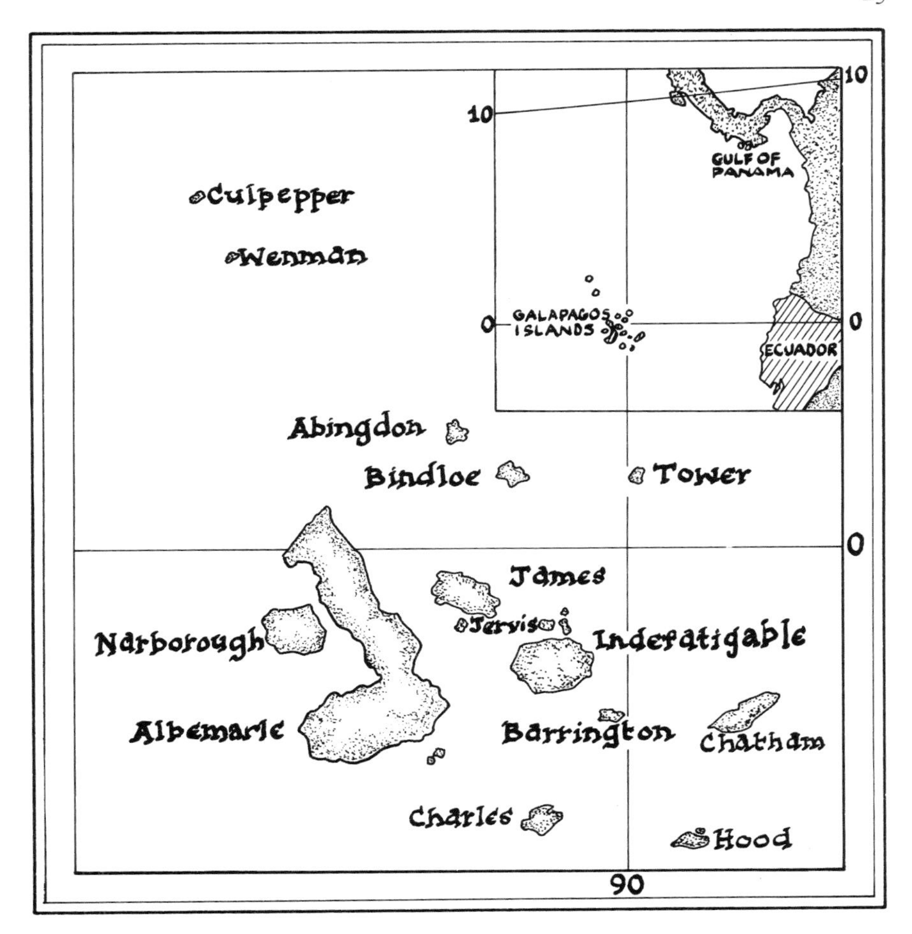


**Fig 6** Map of the Galapagos Islands

ground-feeders taking seeds and insects. Inter-specific competition has been avoided by specialising to some extent on different types of seed, but also in the development of beaks of different lengths and shapes. One species in particular *Geospiza scandens* has evolved a long curved bill ideal for feeding on the flowers of the prickly pear cacti.

Higher up in the island forest zones live the four species of tree finches with their huge, powerful, hawfinch-like beaks. The diet as you would expect consists mainly of seeds although some insects are taken. There is only one warbler, *Dendroica petechia*, on the Galapagos and this lack of warblers has left a vacant niche which the finches have not failed to exploit. The warbler finch *Certhidea olivacea* has a slender bill and its diet is purely insectivorous, some prey being taken on the wing in the manner of a

Large ground finch
*Geospiza magnirostris*

Medium ground finch
*Geospiza fortis*
Small-beaked and large-beaked individuals from the same island

Sharp-billed ground finch
*Geospiza difficilis*

Small ground finch
*Geospiza fuliginosa*

Cactus ground finch
*Geospiza scandens*

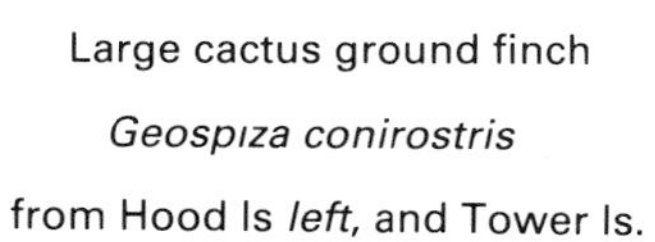

Large cactus ground finch
*Geospiza conirostris*
from Hood Is *left*, and Tower Is.

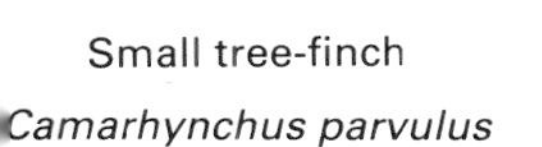

Small tree-finch
*Camarhynchus parvulus*

Medium tree-finch
*Camarhynchus pauper*

Large tree-finch
*Camarhynchus psittacula*
from Charles Is.

Woodpecker finch
*Camarhynchus pallidus*

Mangrove finch
*Camarhynchus heliobates*

Warbler finch
*Certhidea olivacea*

Vegetarian finch
*Camarhynchus crassirostris*

**Fig 7** The Galapagos finches, showing the variety of bill shapes, a phenomenon which was used by Darwin in formulating his evolutionary theory

European flycatcher. The soft fruits and buds are taken by the vegetarian tree finches. Another group of birds absent from the Galapagos are the woodpeckers, and again the resident finches have evolved to fill this niche. The way this problem has been solved is quite remarkable, and unique in the avian world; the finches have no powerful pointed beaks and neither do they have the long probing tongues so typical of the woodpeckers. Instead the finches use cactus spines and twigs to prise insects out of their crevices. One of the most efficient of the woodpecker finches is the mangrove finch *Camarhynchus heliobates*.

These various species of finch have arisen perhaps even from a single species in the following way. A flock of finches must have reached the islands and as time went on the groups interbred and small differences began to develop; this is how sub-species arise. Should those differences widen, the time may come when for one reason or another the different forms no longer interbreed—here we have the birth of a new species. This is just what has happened with Darwin's finches on the islands of the Galapagos.

Britain is an island off the west coast of Europe and some evidence exists to support the theory that Britain has its own sub-species; the pied wagtail, dipper and red grouse all support this contention, and a study of the birds of Britain's offshore islands adds more strength to the argument.

### *The Offshore Islands of Britain*

Britain's offshore islands have not so far given rise to any new species of bird but there are quite a number of sub-specific forms. For example there is a mainland wren, with different sub-species recorded from Shetland, the Hebrides and St. Kilda. Let us return to the island of St. Kilda, once the home of the extinct garefowl, to look at a developing species of wren.

### *The St. Kilda Wren*

The historical references to the wren on St. Kilda go back to at least 1687 when Martin Martin mentions its existence. In 1758 Kenneth McAuley not only recorded it but was even prepared to speculate: 'how could these little birds have flown thither?' It is not until 1829 that it is mentioned again, this time by the resident minister Neil MacKenzie. A naturalist called G.C. Alkinson visited Hirta in 1831 and mentions the wren, but very little light was focused upon its taxonomy until the work of the ornithologist A.G. More was published in 1881. More never actually visited St. Kilda, but he was profoundly interested in the geographical distribution of birds. He suspected that the St. Kilda wren might well prove to be different from those on the mainland. He asked his botanist friend Richard Babbington who visited the islands in June 1883 to obtain specimens for comparison. Barrington found the wren either difficult to catch or so rare that he was not able to deliver the goods. I wonder if Barrington was either a staunch

**Fig 8** St. Kilda, once home of the great auk

conservationist or not quite such a good friend as More imagined.

June 1884 saw one Charles Dixon setting foot on the main island of Hirta, and he found the wren common and was able to write detailed habitat and anatomical descriptions in addition to procuring a number of specimens. Henry Seebohm was able to use these specimens to describe a completely new species which he named *Troglodytes hirtensis*. As we shall see later the St. Kilda wren is now considered as a sub-species, but I must not run on ahead of myself, for the imagined discovery of a new species almost resulted in its extinction as man's thoughts of profit followed their predictable course. By 1894 the people of St. Kilda were doing a roaring trade selling wrens and their eggs to dealers, but the raw material was already beginning to dry up. Indeed W.H. Hudson went so far as to report that, 'the St. Kilda wren no longer exists'. However he was a bit pessimistic with his obituary, and in 1896 Richard Kearton studied and photographed the bird, and in 1902 Dr. Wiglesworth also managed to observe *T. hirtensis* without much difficulty. However the issue was still of such delicacy that parliament stepped in, and Sir Herbert Maxwell had an influential part to play during 1904 in pushing through the houses of parliament The Wild Bird's Protection (St. Kilda) Amendment Act.

As is often the case, however, passing a law is one thing, but applying it is quite another. A dealer called Harry Brazenor must have been confident that

**Fig 9** Village cleitans on St. Kilda, a favourite nesting site for the wren *Troglodytes troglodytes hirtensis*

the law would not be applied since he visited St. Kilda in 1907 and actually advertised a trip in 1908 to collect both the birds and eggs which he said were 'likely to greatly increase in value owing to their growing scarcity and to the increasing stringency of the wild bird protection laws'. So much for the observance of the law in the early years of this century. In the years 1910 and 1911 when the famous naturalist W. Eagle-Clarke visited St. Kilda he reported wrens in the village living in the cleitans and walls, and a substantial population on the cliffs.

An Oxford-Cambridge expedition of the early 1930s set aside two of their team to concentrate upon the distribution and behaviour of the wren and came up with the following data:

| | |
|---|---|
| Hirta | 45 pairs (12 in the village) |
| Dun | 9 pairs |
| Soay | 9 pairs |
| Boreray | 3 pairs at least |
| Minimum Total | 66 pairs |

The village population was counted by Atkinson in 1938 and in 1939 by Fisher, Huxley, Nicholson and Blacker. It is quite remarkable that they all recorded precisely 12 pairs.

In 1947 James Fisher again visited the village, and without being particularly thorough he recorded 10 pairs, so it seems reasonable to assume that he may have missed a couple, and we can assume a fairly steady population within the village. The booklet *Birds of St. Kilda* published by the Institute of Terrestrial Ecology traces population trends, and reports quite a dramatic increase. Whether this is an actual increase, or a reflection of better censusing techniques I cannot say, but by 1957 116 pairs were recorded on Hirta, and the total St. Kildan count was a very healthy 230 pairs. In 1960 there were recorded well over 100 pairs on Hirta, but this had declined to 92 by 1962. In 1957 the wren was reported to be breeding on Stac an Armin (3 or 4 pairs) but so far it has not been recorded from either Stac Lee or Levenish (*see* Fig. 5).

In 1977 six pairs were found breeding in the village but it was stated that the odd pair may have been overlooked. During my visit to the island in 1979 I found eleven nests, but only seven were feather-lined and the rest may therefore have been the cock's nests. Three nests still had young in them, however, and the population would seem to be healthy.

The origin both of the mainland wren and the sub-species unique to these islands is fascinating indeed. Taxonomists are in no doubt that the wren moved to Europe from North America. The bulk of the wanderers seem to have made the complete journey without incident, but some were left isolated on islands as the melt-waters submerged low-lying land. The work done by Charles Darwin on the finches of the Galapagos Islands showed clearly how interbreeding isolated populations adapt to their differing environments by developing often tiny but nevertheless significant differences from the norm. Useful adaptations assist the species whilst those which are not so useful tend towards extinction. Taxonomists feel, at the moment at least, that the St. Kilda wren has not changed sufficiently to be called a separate species, but is different enough to be accorded sub-specific rank. It is therefore called *Troglodytes troglodytes hirtensis* to distinguish it from the mainland wren *Troglodytes troglodytes troglodytes*.

Finally let me consider the physical differences which distinguish the St. Kilda wrens from their mainland relatives, and postulate possible reasons for these differences. Sometime, possibly before the last Ice Age when the climate was milder, the area was forested and possibly not yet isolated, and the wrens colonised St. Kilda. As the climate began to deteriorate and the environment changed the wrens had to adapt to survive. Trees gradually disappeared and the wrens had to learn to perch on the rocks and cliffs.

Three adaptations can be detected in the anatomy of the St. Kilda wren (*see* Fig. 10). The feet are larger and the claws stronger than the mainland form, an obvious advantage when moving about on cliffs in a high wind. The bill is also slightly longer and more powerful, thus assisting the wren to probe deeply into crevices in search of insects. Thirdly the St. Kilda birds are slightly larger than the mainland wrens. This is to be expected, since the larger the body, the less the surface to volume area will be and the less heat will be lost to the environment. Thus in these exposed areas the tendency will be for the wrens of slightly larger size to have a greater chance of survival. Here, then, two of the factors affecting the evolution of new species are working together—geographical isolation and natural selection.

**Fig 10** The St. Kilda wren *Troglodytes troglodytes hirtensis*

*Ring Species*

There is, however, another less direct way in which geographical isolation can produce new species without complete physical separation. This is clearly shown by what have become known as ring species; one example concerns the herring gull *Larus argentatus* and the lesser black-backed gull *Larus fuscus*. In Britain and over a wide area of north western Europe the two species are present and clearly different (*see* Fig. 11) and also have contrasting behaviour and migration patterns. In the wild they very rarely interbreed, but under artificial conditions they prove to be mutually fertile. If you move westwards from the British Isles the lesser black-backed gull becomes less and less common, being almost totally absent from America. On the other hand, the further east you go the lesser black-backed becomes more and more common, and it is the herring gull which disappears. In Siberia a gull is found which is intermediate in appearance between the two; its mantle is lighter than *Larus fuscus* but darker than *Larus argentatus*. This population is therefore assumed to be the one from which *argentatus* and *fuscus* have diverged as they spread one westwards and the other eastwards, thus producing what is called a ring species.

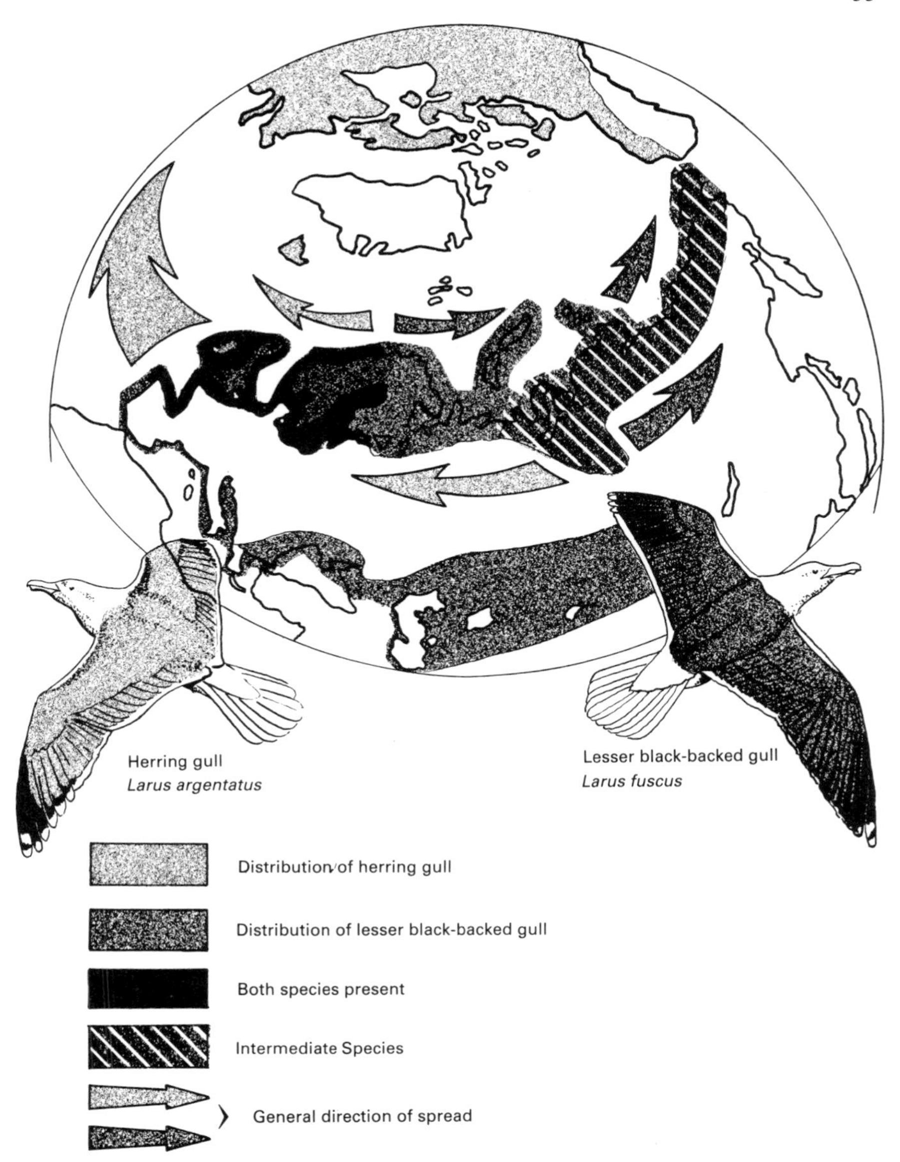


**Fig 11** The map shows the spread of two species of gull to produce a 'super' or 'ring' species

Thus in this chapter we have considered how birds evolved from reptilian ancestors some 150 million years ago, and shown that this evolution continues today. We can now go on in Chapter 2 to show how the 8,700 or so species of bird alive today can be classified.

*Chapter Two*

# The Classification of Birds

Any organised collection of information requires its own rules of nomenclature, and the classification of living organisms is no exception. The system which is in use today is referred to as the binomial system and was the brain child of the eighteenth century Swedish naturalist Linnaeus. Each species is given two names and this combination must be unique to the particular organism. The first name refers to its genus and this is written with a capital letter. *Larus*, for example, signifies that the bird belongs to a genus defining a particular group of gulls. The second name, written with a small letter, refers to the species. Thus the herring gull is *Larus argentatus*, the lesser black-backed gull *Larus fuscus*, the Iceland gull *Larus glaucoides*, the black-headed gull *Larus ridibundus* and so on. By convention, both the generic and specific names are printed in italics.

Moving up the scale, several genera are grouped into a family, and by convention the family name for animals always ends in '-idae'. The family to which the gulls belong is referred to as Laridae. Above the family level there is the order, and similar families are placed into the same order. By convention, the name of an animal order should always end with '-iformes'. Gulls and their relatives, including auks and waders, are placed in a large order called Charadriiformes. Above the level of the order we have the class, above this the sub-phylum, and finally a phylum. Birds belong to the phylum chordata, the sub-phylum vertebrata and to the class birds. The case for the black-headed gull is summarised below:

| | |
|---|---|
| Phylum | Chordata |
| Sub-phylum | Vertebrata |
| Class | Aves (birds) |
| Order | Charadriiformes |
| Family | Laridae |
| Genus | *Larus* |
| Species | *Larus ridibundus* |

It should be noted that the scientific names are often just as descriptive as the vernacular ones, *ridibundus*, for example, meaning 'laughing', and obviously referring to the call-note of the species. It is amusing and often very informative to unravel the meaning of the scientific name.

We can now go on to see how the class Aves is classified. There is some dispute, but most authorities accept the position shown in Table 3. This postulates 33 orders, but 5 of these are extinct, leaving 28 living orders to be described. This scheme shows the Tinamiformes as the most primitive

Table 3
THE CLASSIFICATION OF BIRDS

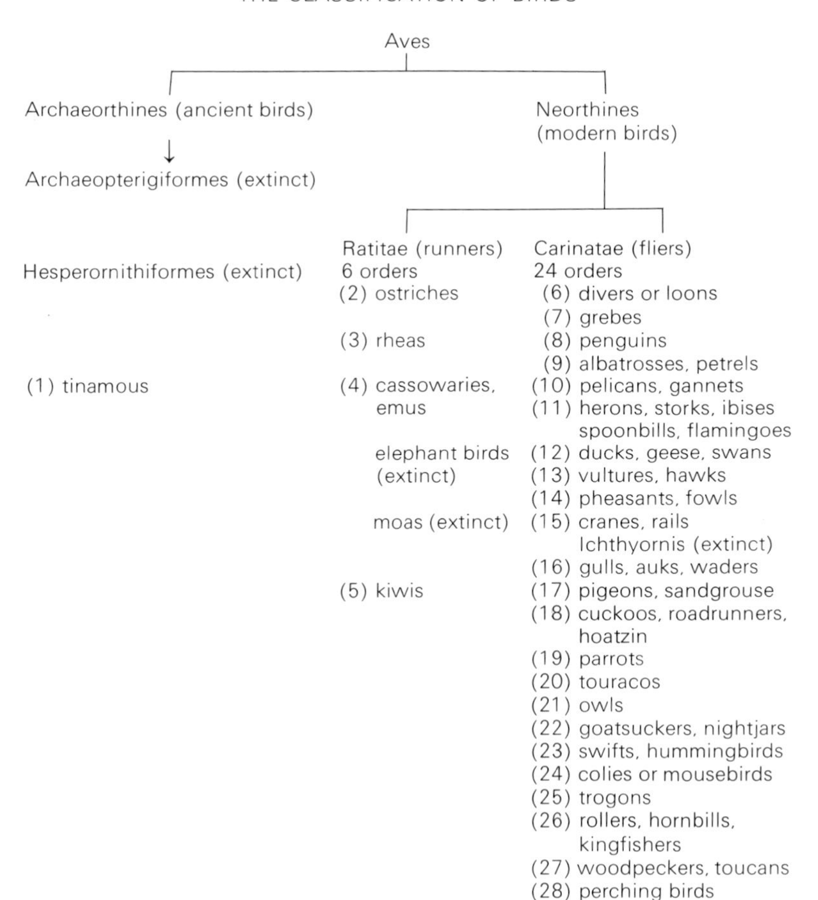


*Note only the living orders are numbered*

order and the Passeriformes or perching birds as the most advanced. I am well aware that between these extremes not everyone accepts the sequence in which I have listed the orders, but no scheme is at present totally accepted by all workers in the field.

With the limited space available I am more conscious of what I have left out rather than what I have managed to include, but the aim is to provide a framework, and at the end of the book an extensive bibliography is given so that anyone interested in a particular order will at least have a starting point for further investigation.

*Order 1*

**(Tinamous)**

## TINAMIFORMES

This primitive order is restricted to Central and South America, an area technically referred to as the Neotropical zone (*see* Chapter 11). An attempt was made to introduce the rufous tinamou to Brightingsea, Essex in 1883, but all had died out by 1888.

Some of their anatomical characteristics are similar to the ratites, but they do have some sort of keel which serves as a point of attachment for flight muscles; the flight, however, is not strong and it should be noted that some authorities place the tinamous between the kiwis and the penguins.

**Fig 12** Rufous tinamou *Rhynchotus rufescens*

The order is divided into nine genera: these have not been studied in sufficient depth to allow a firm decision to be made regarding the precise number of species, but round about 50 would seem to be an acceptable estimate. Tinamous are all cryptically coloured, very grouse-like with short tails, but with de-curved bills. Their call, especially during the breeding season, is described as a beautiful whistle. The sexual roles are reversed, the smaller male being responsible for the incubation of the unicoloured eggs which have a hard, shiny and very attractive appearance. There are records of two females laying their eggs in one nest and then leaving a single male to attend to the family chores.

The diet is almost totally vegetable, as indicated by the very large crop and caeca, the significance of which is described in Chapter 6.

*Order 2*
## (Ostriches)
## STRUTHIONIFORMES

There is only one living species (*Struthio camelus*) and the male, which is larger than the female, can reach a height of 2.5 m (8 ft) and weigh over 150 kg (over 340 lb). It has, however, been divided into five sub-species, one of which, *Struthio camelus syriacus*, seems to have been extinct since about 1940. The reason for this extinction was undoubtedly because it was hunted both for sport and for its feathers, the two factors which have been responsible for huge declines in the populations of the other sub-species. At one time ostriches were common in Africa and south-east Asia, but are now found truly wild only in parts of East Africa. They have, however, been farmed in South Africa since the middle of the nineteenth century and have been introduced to the USA, Europe and Australia, where they breed in captivity. The skin of the bird is still used to produce a first rate leather, but the millinery trade no longer needs the feathers.

The breast bone of the ostrich lacks a keel and so it is totally incapable of flight but it can run at speeds approaching 65 k.p.h. (40 m.p.h) and as in running mammals there has been a reduction in the number of digits in the feet, only the second and the third, much larger, digit remaining.

**Fig 13** Ostrich *Struthio camelus*

*Order 3*

## (Rheas) RHEIFORMES

The rheas are really the South American equivalent of the ostrich, being ratites, adapted for running rather than for flight, and, typically of the ratites, lacking the keel. They lack a hallux (hind toe), but have apparently not travelled so far along evolution's terrestrial road as the ostriches, since rheas have three toes compared to the ostrich's two.

Altogether six species of rhea have been described, and all but two are extinct. These are easy to separate since they differ greatly in size: the greater or common rhea *Rhea americana* stands about 140 cm (4 ft 8 in) compared with the lesser or Darwin's rhea *Pterocnemia pennata* which is only 90 cm (3 ft) high. The two species also have distinct habitat preferences, Darwin's rhea often being found at altitudes of 5,000 m (16,000 ft), whereas the common rhea prefers the much lower pampas lands.

Both species, outside the breeding season, are quite gregarious and flocks of 50 are often encountered. During the breeding season the males collect a harem of several females, all of whom lay in a communal nest sometimes containing 100 eggs to which the male devotes his attentions until the young hatch: this can sometimes be as long as five months.

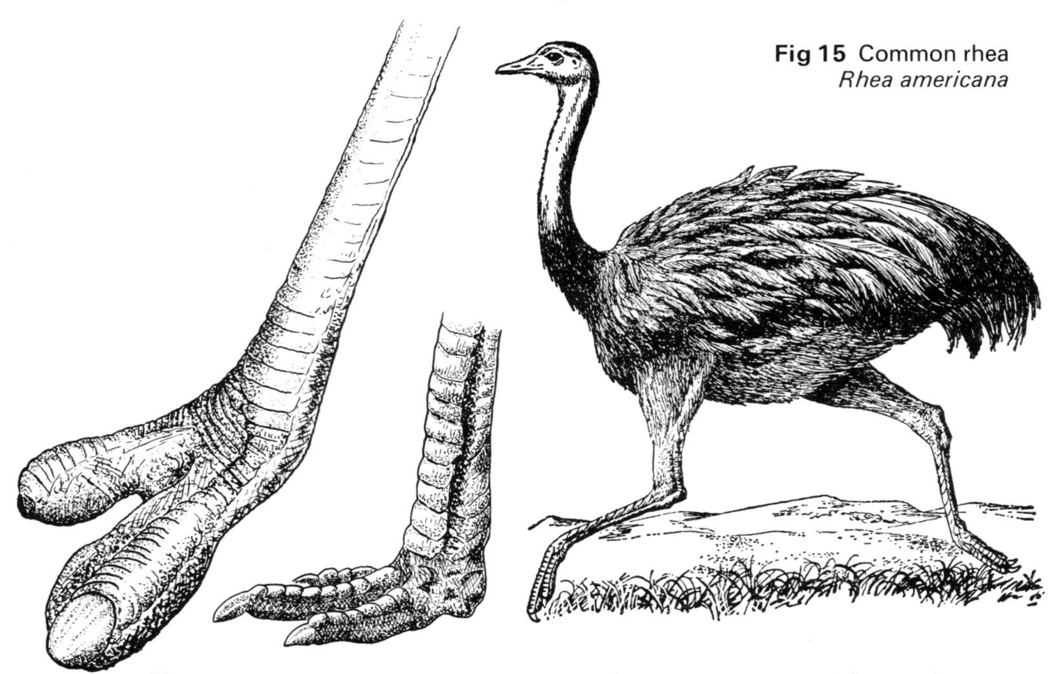

**Fig 15** Common rhea *Rhea americana*

**Fig 14** The rhea and the ostrich both have feet highly developed for running. The rhea is three-toed, and the ostrich two-toed

*Order 4*
## (Cassowaries, Emus)
## CASUARIIFORMES

This order fills the ratite niche in Australia and New Guinea. There is only one species of emu, *Dromaius novae-hollandiae*, which grows to a height of 1.5–1.8 m (5–6 ft) and may weigh as much as 41 kg (96 lb), the females being larger than the males in all respects. The diet is mainly fruit, but invertebrates are taken at times when they are abundant. At the time of the first European settlements along the southerly coastline of Australia three other species occurred, but were soon destroyed and became extinct; these were the Kangaroo Island emu, the King Island emu and the Tasmanian emu.

There are three living cassowaries. The dwarf cassowary *Casuarius bennetti* is found in New Guinea, seldom below 1,000 m (3,500 ft) and stands about 1.1 m (3 ft 6 in) high. The Australian or double wattled cassowary *Casuarius casuarius* can reach a height of 1.8 m (6 ft) and is found in the rain forests of New Guinea. Although mainly fruit-eaters, their huge sharp claws have at times proved capable of killing human beings and they can resort to carnivorous habits if needs be. The 1.6 m (5 ft) single wattled cassowary *Casaurius unappendiculatus* is found along the coastal swamps and riversides of New Guinea and can also be quite a fearsome adversary when cornered.

**Fig 16** Cassowary *Casuarius casuarius*

*Order 5*
(Kiwis)
## APTERYGIFORMES

In many ways the kiwis of New Zealand, almost certainly related to extinct moas, are the most mammal-like of all birds. The long, curved bill has nostrils at the tip rather than at the base, and the sense of smell is quite highly developed. At the base of the bill are specialised feathers which look like hairs and function in the manner of a mammal's whiskers. The wings are vestigial and the feathers are sufficiently soft to feel like fur.

There is some conflict regarding the precise classification within the one recognised genus called *Apteryx*. Most authorities separate the brown kiwis from the spotted kiwis, but others maintain that these differences are external rather than anatomical and they should therefore be regarded as subspecies. There are three forms of brown kiwi, and two spotted types, giving five variations in all.

The kiwi is still abundant in many parts of New Zealand; the females are larger than the males and the roundish body varies from 37–55 cm (15–22 in) giving the bird a superficial resemblance to both a small ostrich and a woodcock thus giving rise to the German name of Schnepfenstrauss—'woodcock ostrich'. A great deal more research is needed before the precise evolutionary history of the kiwis can be unravelled.

**Fig 17** Kiwi *Apteryx australis*

*Order 6*
## (Divers or Loons)
## GAVIIFORMES

This is the first order to be represented in modern day Europe, and is very compact, having only one genus, *Gavia*, comprising four species, all highly specialised for swimming and diving but having adequate powers of flight; on land, however, they are very ungainly. The three front toes are always webbed and the bills are more or less straight and pointed. The species are red-throated *G. stellata*, black-throated *G. arctica*, great northern *G. immer* and white-billed *G. adamsii*. Some workers have listed a fifth species, *G. pacifica*, but most assume this to be a variety of *G. arctica*, the black-throated diver. We must however not be too dogmatic even in such a simple order as the Gaviiformes since it was not very long ago that the white-billed and great northern diver were thought to be varieties of the same species. Superficially they resemble the grebes, but the two orders have followed entirely different evolutionary routes.

Divers invariably lay two eggs which are incubated by both parents for between 25 and 30 days. The chicks leave the nest soon after hatching and fly in about 60 days. They do not breed until their third calendar year.

**Fig 18** Red-throated diver *Gavia stellata*

*Order 7*
## (Grebes)
# PODICEPEDIFORMES

Grebes have a world-wide distribution and are characterised by long straight bills and having their toes lobed rather than webbed. They are highly specialised for an aquatic life and are arranged in one closely knit family, the Podicepedidae, which does not seem to be related to any modern order or indeed to any fossil forms so far discovered. In old books the grebe family was called the Colymbidae. The family divides fairly conveniently into two groups, the distinction being made partly on anatomical differences but also upon behavioural characteristics. There are the pied-billed grebes and the dabchicks, and secondly the ornamental grebes including the Slavonian and great crested grebes. Depending upon the authority consulted the number of species varies from 19 to 22.

Whilst the staple diet for all species consists of invertebrates, the bill shape of various species differs considerably, which helps to reduce harmful competition between them (*see* Chapter 6). The black-necked (eared) grebe *Podiceps nigricollis*, for example, has evolved an upturned bill suitable for scooping up surface insects, whilst the western grebe *Aechmophorus occidentalis* spears fish, and the pied-billed grebe *Podilymbus podiceps* has a short, powerful bill enabling it to crush the hard exoskeleton of crustaceans which make up the bulk of its diet.

**Fig 19** Great crested grebe
*Podiceps cristatus*

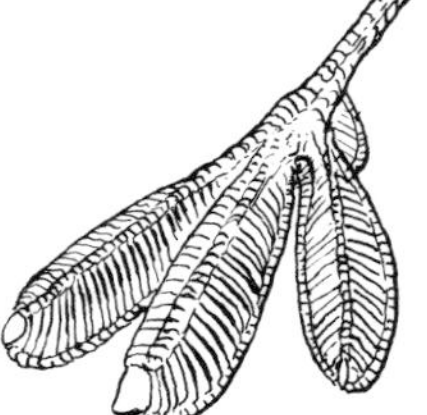

**Fig 20** Lobed toes typical of the grebes

*Order 8*

## (Penguins)
# SPHENISCIFORMES

There are 18 species of penguin alive today and although they are now all flightless there can be no doubt that they have evolved from flying ancestors. The keel of the breastbone is very highly developed and the huge pectoral muscles associated with it power the wings which lack primary feathers, and function as very efficient paddles. The feet are webbed and during movement through water trail behind and serve as a rudder. The toes have strong nails and these are very useful as the animal walks over ice floes, with its upright gait made possible by the fact that the legs are set well back on the body. There is a considerable size variation, the smallest being the Australasian little blue *Eudyptula minor*, weighing only 1–1.2 kg (2–2½ lb), and the largest the 1 m (3 ft) tall Emperor *Aptenodytes forsteri* which has an average weight of about 30 kg (65 lb).

Penguins are restricted to the southern hemisphere and have several important adaptations for living in and around cold water. A layer of blubber beneath the skin cuts down heat loss and there is also a dense overcoat of feathers; those on the body are curved, and overlap like slates on a roof, and beneath these are the down feathers. The combination of the two ensures that the skin remains dry.

**Fig 21** Humboldt penguin *Spheniscus humboldti*

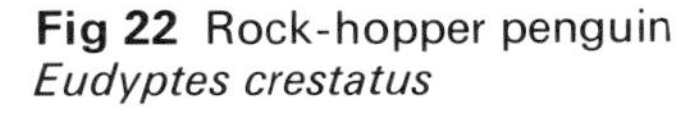

**Fig 22** Rock-hopper penguin *Eudyptes crestatus*

*Order 9*

## (Albatrosses, Petrels)
## PROCELLARIIFORMES

This is a large, complicated order varying a great deal in size from about 25 cm (10 in) to almost 150 cm (60 in). All workers accept the division into four families, namely the albatrosses (Diomedeidae), the shearwaters (Procellariidae), the storm petrels (Hydrobatidae) and the diving petrels (Pelecanoididae). All are typified by having a pronounced hook at the tip of the bill and are unique in possessing the external nostrils modified into tubes and are thus often called the tube-noses. The beak is made up of a series of plates with clearly visible sutures between sections. The three front toes are webbed.

The precise classification within the first three families is pretty well agreed, there being 13 species of albatross, 53 shearwaters, and 20 storm petrels, but the fourth family, the diving petrels, are still causing taxonomic arguments. They look rather like the northern auks (*see* Order 16, Charadriiformes) but are totally confined to the southern hemisphere. A feature of the family is the simultaneous loss of all the flight feathers which means that there is a flightless period during the annual moult. The most recent evidence suggests that the family is made up of four closely related species of the genus *Pelecanoides*: the Peruvian diving petrel *P. garnotii*, the Magellanic diving petrel *P. magellani*, the Georgian diving petrel *P. georgicus* and the common diving petrel *P. urinatrix*.

**Fig 23** Southern giant petrel *Macronectes giganteus*

## *Order 10*
## (Pelicans, Gannets)
# PELECANIFORMES

The Pelecaniformes is an ancient and cosmopolitan order of comparatively large, web-footed, fish-eating birds which at one time were called the Steganopodes. They are in fact the only birds which have all four toes connected by webbing. The order is divided into six families. The tropic or bosun birds (the Phaethontidae) comprise three tropical species. There are six to eight species of pelican (Pelecanidae) depending upon the authority consulted, and they all have huge gular pouches and long broad wings. The boobies and gannets (Sulidae) consists of nine 'types' (three gannet and six boobies) although whether they are all true species or merely of sub-specific rank is open to debate. The cormorants (Phalacrocoracidae) make up the largest family of the order there being 29 or perhaps 30 different species, of which the cormorant *Phalacrocorax carbo* and the shag *Phalacrocorax aristotelis* are north European examples. The tropical darters, also called snake birds (Anhingidae), contain but four species and finally there are the five species of frigatebirds (Fregatidae).

**Fig 24** Shag
*Phalacrocorax aristotelis*

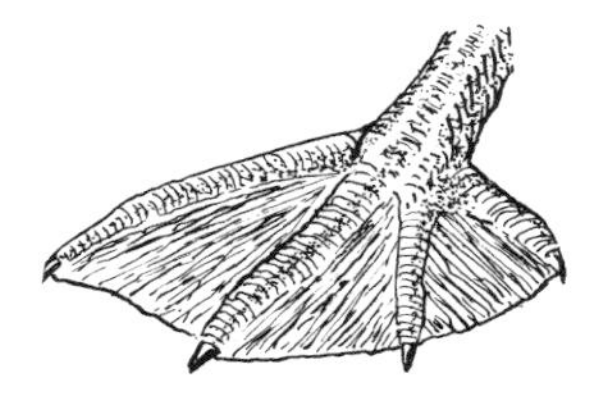

**Fig 25** The cormorant *Phalacrocorax carbo* has broad webbed feet

## *Order 11*
## (Herons, Storks, Ibises, Spoonbills, Flamingoes)
## CICONIIFORMES

At one time called the Ardeiformes, this order contains large, mainly fish-eating birds specialising in wading rather than swimming. Ciconiiformes are divisible into seven families including the 62 species of heron and bittern (Ardeidae), the 17 storks (Ciconiidae), the 31 ibises and spoonbills (Threskiornithidae) and the very interesting flamingoes (Phoenicopteridae). The three minor families are the Cochlearidae consisting only of the boat-billed heron *Cochlearius cochlearius* of Central and South America; the Balaenicipitidae including the whale-headed stork *Balaenicips rex* of East Africa; and the African Scopidae, again made up of one species, the Hammerhead *Scopus umbretta*. These last three families are typified by grotesque, but highly evolved bill shapes.

The flamingoes are a most intriguing family of four or more likely five species which have been placed into three genera, the basis for separation being largely based on bill and leg colour, although the family itself is typified by having very long legs and gracefully curved necks but very short webbed toes. Flamingoes belong to the oldest bird family still alive and fossil evidence places them firmly in the Tertiary period (*see* Chapter 1). These fossils indicate that far from being confined to saline or alkaline lakes in the tropics, flamingoes were once widespread throughout Australasia, America and Europe. The bill of flamingoes has a sieving device very similar to that found in ducks, and many workers consider the Phoenicopteridae to be the link between the Ciconiiformes and the Anseriformes.

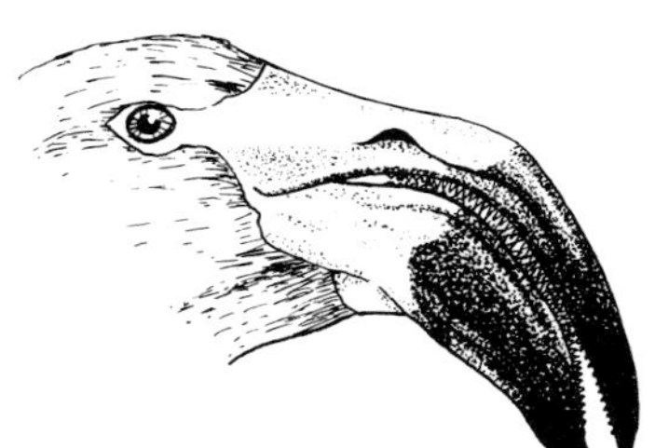

**Fig 26** The bill of the Chilean flamingo *Phoenicopterus chilensis*

**Fig 27** Great blue heron *Ardea herodias*

## *Order 12*
## (Ducks, Geese, Swans)
## ANSERIFORMES

This order includes birds which are found in marshy areas and lay unspotted eggs, and usually have webbed feet. They are usually classified into two families. Firstly there is the Anhimidae, a South American family often referred to as the screamers. Three species are recognised and are distinguished from the waterfowl to which they are anatomically related by having a bill more like that of a gamebird than that of wildfowl: indeed they may well be one of the links between the two. Screamers also moult their flight feathers gradually and therefore do not have the flightless period typical of waterfowl. The legs are longer than those of wildfowl, and the feet have little webbing and have a highly developed hind toe. The second family is the Anatidae or waterfowl and this is made up of 146 species divided into 37 genera. Two sub-families are recognised, namely the Anseranatinae including swans, geese and whistling ducks, and the Anatinae consisting of the burrowing shelducks and the ducks. Finally the ducks themselves are classified into perchers (genus *Cairina*), dabblers (genus *Anas*) divers (genera *Aythya* and *Netta*) sea ducks (genus *Mergus*, the mergansers; and *Somateria*, the eiders) and stifftails (*Oxyura*, the ruddy ducks).

**Fig 28** Black swan *Cygnus atratus*

**Fig 29** A pair of mandarin ducks *Aix galericulata*

*Order 13*
## (Vultures, Hawks)
## FALCONIFORMES

It must be true to say that throughout the world no other order of bird has evoked so much admiration and interest in the human mind than the diurnal birds of prey. This is amply borne out by the volume of literature available. They are classified into five families. The secretary bird *Sagittarius serpentarius* is placed in a family of its own, the Sagittaridae, and is restricted to Africa south of the Sahara. Not all taxonomists are happy to see this 120 cm (4 ft) tall, running bird placed in the Falconiformes, despite both its bill and talons being hawk-like. The second family is the Cathartidae, the New World vultures, consisting of seven species arranged in four genera and including the endangered but still mighty Andean condor *Vultur gryphus*. The family Accipitridae embraces some 217 species of hawks and eagles arranged into 64 genera, and the Falconidae family has 10 genera and 61 species. The falcons and hawks resemble each other in having powerful bills and talons as well as acute eyesight, but are separated on several points. The inside of the eggshell when held up against the light is buff-coloured in falcons and greenish in hawks and eagles. The latter family defaecate by forcibly directing the matter several feet from the nest whilst falcons defaecate directly beneath the perch. The classification of Falconiformes is completed by the Pandionidae, which consists of only one species, the fish-eating osprey *Pandion haliaetus*.

**Fig 30** Kestrel
*Falco tinnunculus*

*Order 14*
## (Fowls, Pheasants)
## GALLIFORMES

This order is typified by the domestic fowl, being adapted for walking rather than flight as the very short rounded wings clearly show. Although the diet is varied, vegetable food is preferred and a strong muscular gizzard aids digestion. The birds range in size from the splendid peacock at one extreme to birds of sparrow size at the other. The order is divisible into six families. The Megapodiidae (megapodes) comprises five species of mallee fowl from Australia and Polynesia and the six species of Australian brush turkeys. The Cracidae (guans) is a family of 44 species arranged in 11 genera, mostly restricted to South America, but fossils have been found in France, Patagonia and in the north and east of the U.S.A. The Tetraonidae (the grouse) comprises between 11 and 17 species depending upon the authority consulted. One species, the red or willow grouse *Lagopus lagopus*, has become a vital part of the economy of northern Britain, all other forms of natural life being sacrificed to maintain the glory of 12 August, the opening of the grouse season. The Phasianidae (pheasants and quails) with its 183 species is probably of greater importance than any other family of birds since it includes the domestic fowl which has been associated with mankind for thousands of years. Also often kept for eating are the Numididae (Guinea fowl) of which there are eight species arranged in four genera. The sixth family is the Meleagrididae (the turkeys) of which only two species are found, the common *Meleagris gallopavo* and the ocellated turkey *Agriocharis ocellata*. They, too, have been hunted for food and were taken to Europe from the Americas in the sixteenth century by the Spanish. They acquired the name 'turkey' because this was a term used at the time to describe any object from abroad.

**Fig 31** Peacock *Pavo cristatus*

## *Order 15*
## (Cranes, Rails)
# GRUIFORMES

Formerly known as the Ralliformes, this cosmopolitan order is divided into 12 families, roatelos, button quails, the plains wanderer, the limpkin, trumpeters, sun grebes, sun bitterns, seriemas, the kagu and the most significant families of the bustards, cranes and rails. The Otididae (the bustard family) is confined to the Old World; its twenty two species are mainly African, with a few reaching Europe and one family being found in Australia. The flesh makes excellent eating, and since the birds prefer to crouch rather than fly when disturbed they are easily killed and therefore face extinction in many areas. The 14 cranes are big long-legged birds and have been present since the Eocene period (*see* Table 1); they are among the world's most efficient migrants, which may well account for their occurrence in Europe, North America, Asia, Africa and Australia. The rails (family Rallidae) comprise about 140 species including the gallinules and the coots, and they are present on all continents except Antarctica. There have been several flightless island forms which have been vulnerable enough to become extinct in comparatively recent times as a result of man's demand for easy sources of pleasant tasting proteins, or, perhaps more significantly, preyed upon to an intolerable degree by cats, dogs, or rats introduced to those islands by man.

**Fig 32** Blue-necked crowned crane *Balearica pavonina regulorum*

**Fig 33** Pectoral rail *Rallus philippensis australis*

## *Order 16*
## (Gulls, Auks, Waders)
## CHARADRIIFORMES

This is a huge order made up of 16 families which can conveniently be described in tabular form.

| Family | Number of species | Comment |
|---|---|---|
| Jacanidae—jacanas | 7 perhaps 8 | Unwebbed feet with long toes. Walks on lily leaves. |
| Rostratulidae—painted snipe | 2 | *Rostratula benghalenses* (Old World); *Nycticryphes semicollaris* (S. America). |
| Haematopodidae—oystercatchers | 4 (21 sub-species) | Specialist feeders on bivalve molluscs. |
| Charadriidae—plovers | 63 | Cosmopolitan except for Antarctica. |
| Scolopacidae—sandpipers | 82 | Mainly shore birds. |
| Recurvirostridae—avocets, stilts | 7 | Avocets, long curved bills; stilts, long legs, straight bills. |
| Phalaropodidae—phalaropes | 3 | Females brighter coloured than males. Sexual roles reversed. |
| Dromadidae—crab plover | 1 | Only one species *Dromas ardeola*. Hard to place. 'cont.' |

**Fig 34** Oystercatcher *Haematopus ostralegus*

| Family | Number of species | Comment |
|---|---|---|
| Burhinidae—thick knees or stone curlews | 9 | 2 sub-families: the *Esacus* (2 species confined to Australia); *Burhinus* (7 species including stone curlew *Burhinus oedicnemus*). |
| Glareolidae—pratincoles, coursers | 17 | 2 sub-families: the pratincoles (Glareolinae) 9 species; coursers (Cursoriinae) 8 species. |
| Thinocoridae—seed snipes | 4 | Confined to S. Africa. |
| Chionididae—sheathbills | 2 | Found around Indian Ocean and Antarctic. |
| Stercorariidae—skuas, jaegers | 6 | 3 skuas and 3 jaegers, but classification confused at the moment. |
| Laridae—gulls, terns | 86 | 45 gulls, 41 terns. |
| Rynchopidae—skimmers | 3 | The three are only just recognisable as separate species being very closely related. |
| Alcidae—auks | 24 | A very ancient order including the great auk (*see* Chapter 1). Replace the penguins in the Northern Hemisphere. |

**Fig 35** Lapwing *Vanellus vanellus*

**Fig 36** Kittiwake *Rissa tridactyla*

*Order 17*

## (Pigeons, Sandgrouse)
## COLUMBIFORMES

The extinct, flightless dodo of Mauritius was related to the pigeons and placed in the family Raphidae; the other two families constituting this order are the Pteroclidae or sandgrouse (16 species) and the pigeons themselves (the Columbidae). The sandgrouse, as their name suggests, are very grouse-like in appearance, but they live in dry open country, have small feet, and in flight their wings are seen to be long and pointed. The Columbidae have been associated with human settlements for thousands of years and some 255 species are recognised; although the smaller members are often called doves, in contrast to the larger pigeons, the two terms are synonymous and have no valid taxonomic distinction. There are few orders more successful than the Columbiformes and within it have evolved species of varying shapes, sizes and colourations, a performance known technically as adaptive radiation, and which drew the great Charles Darwin into a detailed study of their nature. The Columbidae have been useful to man for food and, because of their strong 'homing' instinct, to carry messages, but because of their dependence on crop plants for food many species are considered as economic pests.

**Fig 37** Black-bellied sandgrouse
*Pterocles orientalis*

**Fig 38** Wallace's fruit dove
*Ptilinopus wallacii*

*Order 18*

## (Cuckoos, Roadrunners, Hoatzin)
## CUCULIFORMES

Closely related to the parrots this order is made up of two fascinating families. The Opisthocomidae is made up of one species, the hoatzin *Opisthocomus hoazin*, which is unique in having young with claws upon their wing tips; the species was therefore classified at one time as a very primitive bird closely related to Archaeopteryx. It is now thought that the claws have evolved more recently as an adaptation to enable the young bird to hang on to the swaying branches of trees before it can fly. The claws are absent in the adult bird. The Cuculidae or cuckoo family is, despite views expressed by early ornithologists, very closely knit, but does show considerable diversity in breeding biology. Nest parasitism, so much a feature of the European cuckoo *Cuculus canorus*, is found only in a minority of the 127 species and is mainly restricted to one of the six sub-families, the Cuculinae. Also belonging to the cuckoo family are the peculiar looking roadrunners.

**Fig 39** Hoatzin *Opisthocomus hoazin*

*Order 19*

## (Parrots)

# PSITTACIFORMES

The precise taxonomic position of parrots is subject to some argument, but they do have some connection with the cuckoos and perhaps even more so with the pigeons. The 330 species are, however, regarded as being so similar as to be placed all within one family, the Psittacidae. They are mainly restricted to tropical areas such as South and Central America, and the hot areas of Australasia and southern Asia. Africa has only 15 species and Europe and North America have no native species although the extinct Carolina parakeet *Conuropsis carolinensis* was formerly present at least as far north as the Great Lakes.

It is, of course, as aviary birds that the seed- and fruit-eating parrots, cockatoos, lories, lovebirds and macaws are best known. The budgerigar *Melopsittacus undulatus* (in the past exported in great numbers from Australia) is perhaps the best known bird, but the cockatiel *Nymphicus hollandicus* and the African grey parrot *Psittacus erithacus* are also popular birds with aviculturalists. All have a powerful hooked bill and grasping feet with two toes pointing forwards and two backwards (called zygodactyl) ideal for tree living, seed- and fruit-eating birds. Many are colonial, and this is particularly true of budgerigars which have been observed breeding in groups of nearly a thousand pairs, and flocks approaching one million birds have been seen in the wild.

**Fig 40** Lesser Patagonian conure *Cyanoliseus patagonus*

*Order 20*

## (Touracos)
## MUSOPHAGIFORMES

There are 22 species within the one family, Musophagidae, and these vary in size from 37–71 cm (14½–28 in). They are found in Africa south of the Sahara, but they do not occur in Madagascar.

They are basically fruit- and insect-eaters and they catch the insects in trees, preferring rather to run along branches and to hop from tree to tree than to fly. The plumage is very attractive indeed, including glossy greens, reds, violets and yellows. The stout bill is also often brightly coloured and most birds are crested. Females and males look alike, and the family tends to be sedentary. An interesting feature of the family is that the fourth toe is reversible, a condition known as semi-zygodactyl.

**Fig 41** Black-billed touraco *Tauraco schüttii*

*Order 21*

## (Owls)
## STRIGIFORMES

The earliest fossil owl so far discovered has been named *Protostrix mimica*, and it was found close to Ten Mile Creek in Wyoming, U.S.A. This has been dated at some 60 million years, which places it in the Palaeocene period (*see* Table 1). Owls are divided into two families, the Strigidae, made up of about 120 species, and the Tytonidae made up of about 11 species. The Strigidae (eagle and eared owls) is by far the oldest family and fossils found in France have been placed in the Oligocene period (36 million years old) whilst the Tytonidae (barn owls) are much younger and the earliest fossils so far unearthed date back only 12 million years to the Miocene period.

Owls are basically nocturnal birds of prey, the dark hours' equivalent of the diurnal hawks, but to which they bear no taxonomic resemblance. The disc-like feathering in the region of the eyes give them a distinctive appearance, and the soft fluffy feathers give them a roundish shape, but, more importantly, enable them to fly almost silently, a useful attribute for a creature of the night. The position of the eyes gives a certain degree of stereoscopic vision enabling a better judgement of distance than is normal in birds.

**Fig 42** Tawny owl *Strix aluco*

**Fig 43** Barn owl *Tyto alba*

*Order 22*
## (Goatsuckers, Nightjars)
## CAPRIMULGIFORMES

The order is made up of five families. The oilbirds (Steatornithidae) are represented by only one living species, the *Steatornis caripensis* of South America as discovered by Humboldt only in 1799; since then its oil has been so exploited that it has now been given special protection. The frogmouths (Podargidae) are found in south-east Asia and Australia and comprise 12 species. The potoos (Nictibiidae) consist of only five species restricted to the American tropics. There are seven species of nocturnal owlet frogmouths (Aegothelidae) found only in Australia and New Guinea.

All members of the order tend to be nocturnal and resemble the owls in having huge eyes and a brownish grey plumage blotched and barred, giving perfect camouflage. They show some variation in bill structure, that of the fruit-eating oilbird being hooked, whilst those of the potoos and frogmouths also tend to be hooked, but have a wide gape capable of capturing insects on and around trees. In the fifth family, the true nightjars (Caprimulgidae), made up of about 70 species, we find the bills to be much weaker but with a huge gape working like a butterfly net as the bird glides around sweeping up insects. In the European nightjar *Caprimulgus europeaus* the margin of the bill has evolved long bristles to increase the feeding efficiency. The nightjars also have longer, more pointed wings than the rest of the order.

**Fig 44** Tawny frogmouth *Podargus strigoides*

*Order 23*

## (Swifts, Hummingbirds)
## APODIFORMES

Swifts are without doubt the most aerial of birds, and the bill has become reduced to little more than a rim surrounding a mouth which is essentially an insect catching bag. The wings are very long and pointed, and the feet which so seldom come into contact with solid ground are tiny and almost vestigial, the birds even managing to mate and sleep on the wing. The order has three families, the true swifts (Apodidae), the crested swifts (Hemiprocnidae), and the hummingbirds (Trochilidae). The true swifts have caused no little argument among taxonomists, and between 70 and 80 species are recognised depending upon the authority consulted. The crested swifts are a much easier proposition, consisting only of three species placed in a single genus, *Hemiprocne*. Some workers have suggested that the hummingbirds should be placed in an order of their own (the Trochiliformes) but despite being tempted to follow this idea I have left them in their conventional spot; in any case they are very closely related to the swifts. There are 319 species restricted to the New World, the southerly limit being Tierra del Fuego whilst Labrador and Alaska seem as far north as they have been able to penetrate.

**Fig 45** Emerald hummingbird *Amazilia tobaci* on nest

## *Order 24*
## (Colies or Mouse Birds)
## COLIIFORMES

Birds of this order are small fruit-eating birds restricted to the African continent where their vegetable diet can often be a serious menace to the human economy. They are typified by having very long, pointed tails consisting of 10 feathers often twice as long as the body itself, very short legs and the peculiar property of being able to turn every one of their four toes forwards. There are only six species included in a single family called the Coliidae, and the relationship between them is even close enough for them to be included in the one genus *Colius*. Mouse birds are all gregarious, and a large flock can visit a fruit or vegetable crop and reduce it to leafless branches in quick time making them much feared and persecuted in consequence. The range of the genus within Africa south of the Sahara is extensive, ranging from mountain tops of 2,500 m (over 8,000 ft) right down to sea level and as they are non-migratory, some areas are more prone to attack than others.

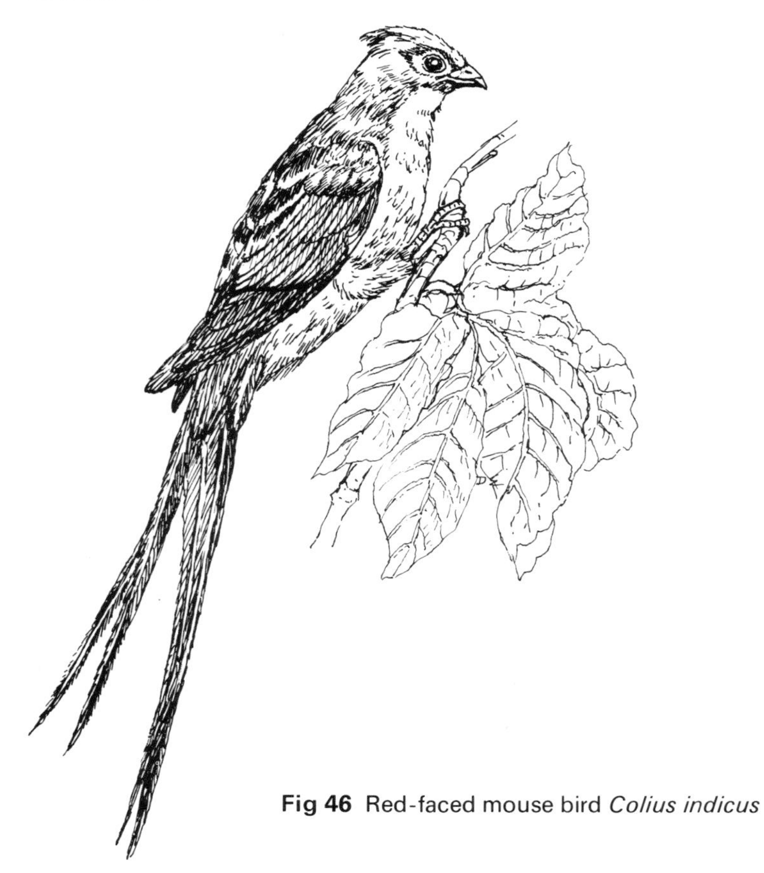

**Fig 46** Red-faced mouse bird *Colius indicus*

*Order 25*

**(Trogons)**

## TROGONIFORMES

The trogons form an order of pan-tropical, medium sized, often exquisitely coloured birds. They have rather long tails and short legs with the inner toe turned backwards to produce a perching foot of two forward and two backward facing toes. The order has proved difficult to place since their relationship with other orders seems to be very obscure. The bill is short, but quite sturdy, and has a surprisingly wide gape, an ideal arrangement for birds which feed on fruit and insects. Fossil trogons have been discovered in southern France which would suggest that in Eocene and Oligocene periods when the climate was much warmer this order was quite widely distributed. Only one family, the Trogonidae, occurs and is made up of between 35 and 40 species depending upon the authority consulted. With regard to both the number of species and actual population levels the Americas appear to be the favoured range of the trogons. They are mainly woodland dwellers spending much of their time perching almost motionless on branches.

**Fig 47** White-tailed trogon *Trogon strigilatus*

*Order 26*

## (Rollers, Kingfishers, Hornbills)
## CORACIIFORMES

The birds of this order, apart from the cuckoo roller, have their front three toes more or less joined at the base. There are nine families including the widespread kingfishers (Alcedinidae), 90 species; the todies (Todidae) of the West Indies, 5 species; and the motmots (Momotidae) of tropical America, 8 species. All these have large bills and usually perch upright prior to diving down onto their prey; only a few of the kingfishers actually feed on fish rather than insects. Also included are the 23 species of Old World bee-eaters (Meropidae) which have specialised in dealing with flying insects. The rollers (Coraciidae), of which there are 16 species, are again restricted to the Old World and resemble bee-eaters to some extent, but instead of the long slender bill they have a short stout bill with a broad gape. The cuckoo roller (Leptosomatidae) is a single species (*Leptosomus discolor*) found only on the Comore Islands and Madagascar. The hoopoes (Upupidae) of the Old World are typified by the possession of crests and long, slightly down-curved bills, there being about seven sub-species of the only modern species *Upupa epops*. The wood hoopoes (Phoeniculidae), 8 species, are restricted to Africa, have long, curved bills and are primarily arboreal. Finally the hornbills (Bucerotidae), 44 species of which are recognised, are found in tropical regions of the Old World. They are larger birds with bills of huge proportions, often with an elaborate casque. They are mainly fruit-eaters with broad wings and small feet.

**Fig 48** Hoopoe *Upupa epops*

*Order 27*
## (Woodpeckers, Toucans)
## PICIFORMES

Perhaps the best known of the Piciformes are the woodpeckers (family Picidae) and the toucans (family Ramphastidae) of which there are 38 recorded species, but also included are four other families which are listed in tabular form below.

| Family | Number of species | Comment |
|---|---|---|
| Galbulidae–jacamars | At least 15, but some confusion exists. | Found in the New World tropics. |
| Bucconidae–puffbirds | About 32. | Tropical New World. |
| Capitonidae–barbets | 72–78 depending upon authority consulted. | Tropical forests of Central and South America, Asia, Philippines and Indonesia. |
| Indicatoridae–honeyguides | About 15. | Africa and Asia. The only flying birds to have only 9 primary feathers. |

All the members of the order are tree-dwellers, having two toes pointing forwards and two behind. The woodpeckers are divided into three sub-families: the true woodpeckers (Picinae) consists of 179 species; the wry-necks (Jynginae) of two, one African and one Eurasian; and the tiny piculets of 43, three south-east Asian, 15 African, and 25 tropical American. The Piciformes show such close similarities to the order of perching birds (Passeriformes) that some past workers have included them as a family within this order, and not as a separate order.

**Fig 49** Wryneck
*Jynx torquilla*

*Order 28*

**(Perching Birds)**

## PASSERIFORMES

Here we have a great assemblage of birds containing over half of all living birds species. Their size varies from the tiny honey-eaters, goldcrest and firecrest to the raven which is over 60 cm (2 ft) long. Most of the species have generalised bill, wings, feet and tail and, as their name implies, have three toes in front and one behind, an ideal arrangement for perching. The syrinx is especially well developed and gives them a considerable vocabulary, and thus song birds are often aesthetically important. There are 68 families listed below in my suggested evolutionary order, but again I am aware that my choice of position for a particular family may not satisfy all taxonomists.

**Fig 50** Head and foot of raven *Corvus corax* and, to scale, firecrest *Regulus ignicapillis*, showing the great variation in size among the Passeriformes

| Family | Number of species |
|---|---|
| Eurylaimidae–broadbills | 14 |
| Dendrocolaptidae–woodcreepers | 50 |
| Furnariidae–ovenbirds | 217 |
| Formicariidae–antbirds | 230 |
| Conopophagidae–antpipits | 8 |
| Rhinocryptidae–tapaculos | 29 |
| Cotingidae–cotingas | 79 |
| Pipridae–manakins | 53 |
| Tyrannidae–tyrant flycatchers | 362 |
| Oxyruncidae–sharpbills | 1 |
| Phytotomidae–plantcutters | 3 |
| Pittidae–pittas | 26 |
| Acanthisittidae–New Zealand wrens | 3 |
| Philepittidae–asities | 4 |
| Menuridae–lyrebirds | 2 |
| Atrichornithidae–scrub-birds | 2 |
| Alaudidae–larks | 76 |

| | |
|---|---|
| Hirundinidae–swallows, martins | 74 |
| Motacillidae–wagtails, pipits | 55 |
| Campephagidae–caterpillar birds | 72 |
| Pycnonotidae–bulbuls | 118 |
| Irenidae–fairy bluebirds, leafbirds | 14 |
| Laniidae–shrikes | 79 |
| Vangidae–vangas | 13 |
| Bombycillidae–waxwings | 8 |
| Dulidae–palmchat | 1 |
| Cinclidae–dippers | 4 |
| Troglodytidae–wrens | 59 |
| Mimidae–mockingbirds, thrashers | 30 |
| Prunellidae–hedge sparrows (accentors) | 13 |
| Turdidae–thrushes | 304 |
| Timaliidae–babblers | 252 |
| Sylviidae–Old World warblers | 339 |
| Maluridae–Australian wren warblers | 29 |
| Acanthizidae–Australian warblers | 59 |
| Muscicapidae–Old World flycatchers | 134 |
| Rhipiduridae–fantail flycatchers | 38 |
| Monarchidae–monarch flycatchers | 133 |
| Pachycephalidae–whistlers | 48 |
| Remizidae–penduline tits | 9 |
| Aegithalidae–long-tailed tits | 7 |
| Paridae–titmice | 46 |
| Sittidae–nuthatches | 21 |
| Climacteridae–Australian treecreepers | 6 |
| Certhiidae–typical creepers | 6 |
| Dicaeidae–flowerpeckers | 58 |
| Nectariniidae–sunbirds | 118 |
| Zosteropidae–white-eyes | 79 |
| Ephthianuridae–Australian chats | 5 |
| Meliphagidae–Australian honey-eaters | 169 |
| Emberizidae–buntings, American sparrows | 233 |
| Parulidae–American wood warblers | 120 |
| Drepanididae–Hawaiian honeycreepers | 15 |
| Vireonidae–vireos | 39 |
| Icteridae–American blackbirds, orioles | 92 |
| Fringillidae–chaffinches, linnets | 126 |
| Estrildidae–waxbills | 124 |
| Ploceidae–typical weavers | 150 |
| Sturnidae–starlings | 106 |
| Oriolidae–Old World orioles | 28 |
| Dicruridae–drongos | 20 |
| Callaeidae–wattlebirds | 3 |
| Grallinidae–mudnest-builders | 4 |
| Artamidae–wood-swallows | 10 |
| Craticidae–bell-magpies | 11 |
| Ptilonorhynchidae–bowerbirds | 17 |
| Paradisaeidae–birds of paradise | 40 |
| Corvidae–crows, jays, magpies | 103 |

It may well appear to some readers that I have dismissed this very important order in a somewhat cursory manner, but those interested in pursuing the subject may redress the balance by reference to the bibliography.

*Chapter Three*

# The Bird's Body

Those beginning the study of ornithology are often worried by the specialised vocabulary used to describe the external features of a bird, and fight shy of writing invaluable field notes because they are not quite sure what to call a particular part of the anatomy. For this reason I have set out at the beginning of this chapter to provide a clear diagrammatic description of bird topography before going on to describe the structure of the various types of feather and to give some account of their colouration and moult. Some attention is then focused upon those skeletal features typical of birds before a description of that unique structure, the avian wing.

## Bird Notes

When I first began to take my birdwatching seriously I made several copies of diagrams similar to Figure 51 and stuck one in the front of each field notebook, and I still do this now. It serves to remind me of the proper way to describe any unfamiliar bird, because however competent you may think you have become, your memory may still let you down.

Apart from such details as weather, and the habitat in which you find your bird, the following details should always be noted. Firstly an estimate of size should be given, not by guessing the dimensions but by reference to named birds, and a good guide to a new bird might be 'bigger than a robin but smaller than a starling'. Then, by reference to the topographical diagram the features should be described as fully as possible, beginning at the tip of the bill and working backwards to the end of the tail. It may well be that the bird does not remain still long enough for a complete description to be compiled, in which case details of flight and any call notes or song patterns should be mentioned. Soon you will know your way around the bird and your enjoyment of the hobby will increase, as will your value to other ornithologists.

## The Form and Function of Feathers

Feathers are unique to birds, each consisting of a tapering shaft bearing a flexible vane on each side. The short, basal part of the feather is called the calamus; in cross section it is round, and it is almost hollow. There is an opening at the bottom of the calamus, the lower umbilicus, and it is through this that the blood supply entered the young feather during its short period of growth. When growth is complete the feather is sealed off and, although

**Fig 51** (opp.) Bird Topography. A drawing such as this will help you to remember the correct way to describe any new bird

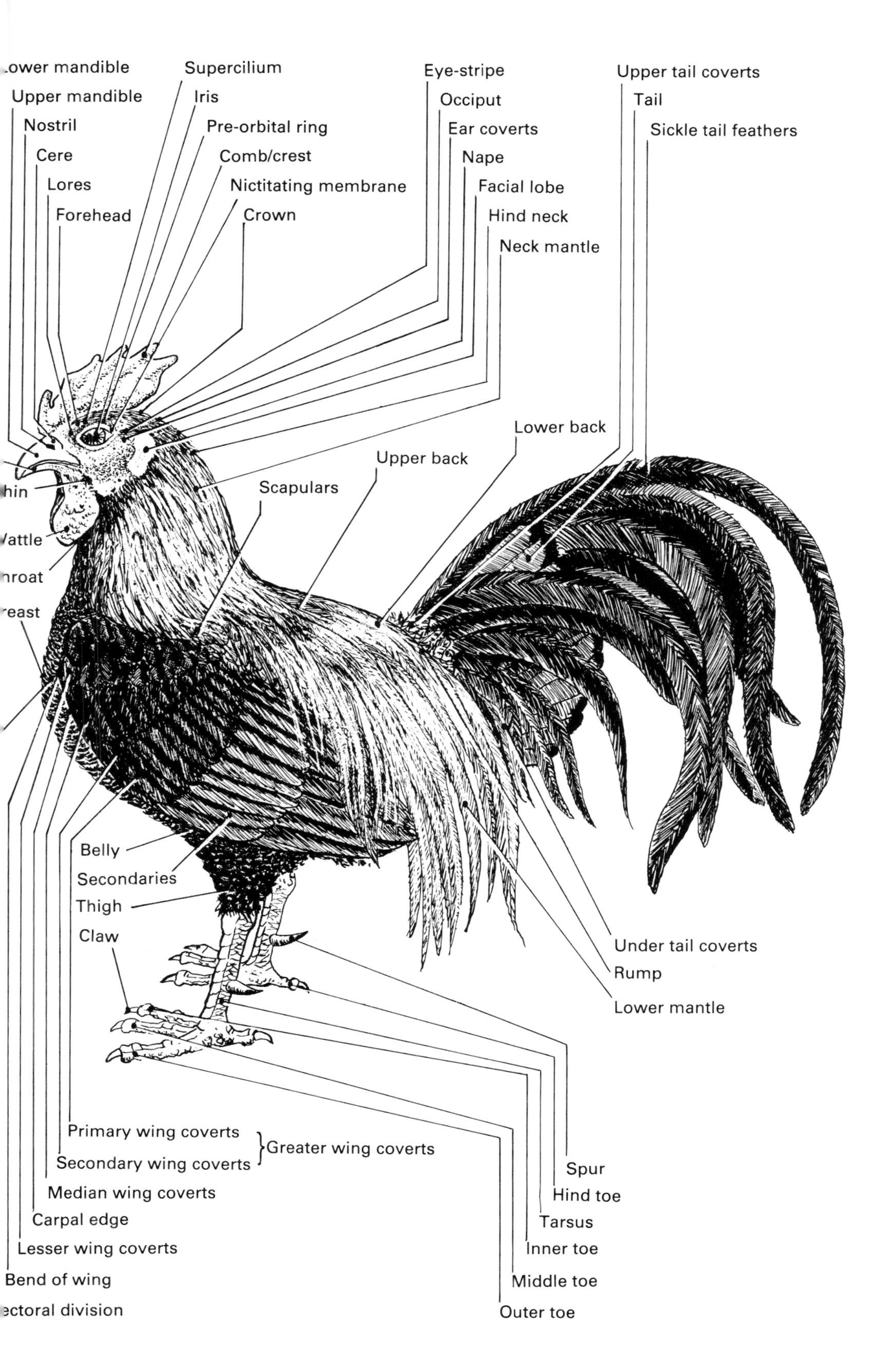

ower mandible
Upper mandible
Nostril
Cere
Lores
Forehead
Supercilium
Iris
Pre-orbital ring
Comb/crest
Nictitating membrane
Crown
Eye-stripe
Occiput
Ear coverts
Nape
Facial lobe
Hind neck
Neck mantle
Upper tail coverts
Tail
Sickle tail feathers
Lower back
Upper back
Scapulars
hin
attle
roat
east
Belly
Secondaries
Thigh
Claw
Under tail coverts
Rump
Lower mantle
Primary wing coverts
Secondary wing coverts
Greater wing coverts
Median wing coverts
Carpal edge
Lesser wing coverts
Bend of wing
ectoral division
Spur
Hind toe
Tarsus
Inner toe
Middle toe
Outer toe

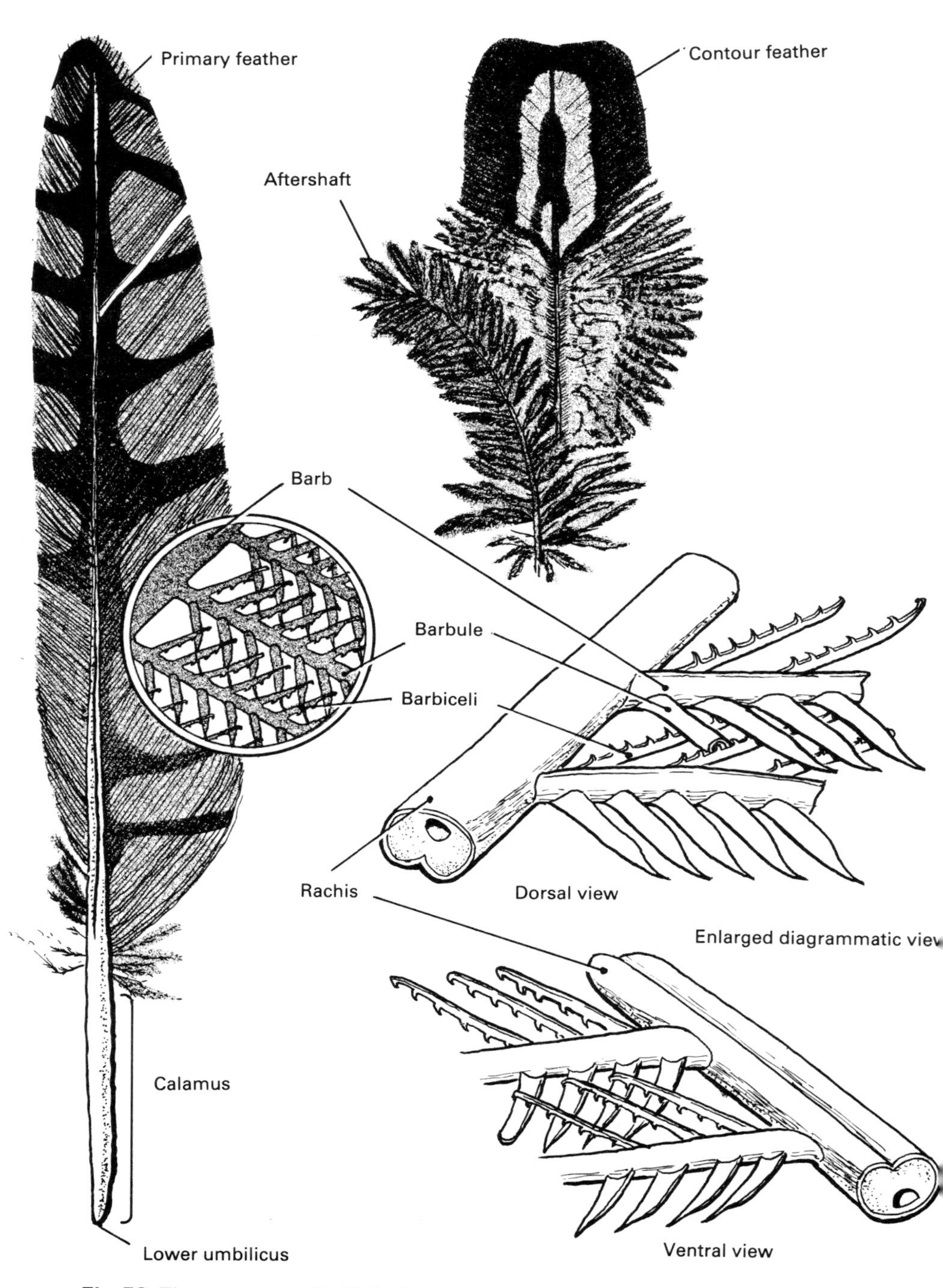


**Fig 52** The structure of a flight feather

each can be moved by a separate muscle situated in the skin, the feather itself is 'dead'. The solid rachis carries the vanes of the feather, and each of these is composed of a row of barbs, arranged side by side and linked together by smaller branches called barbules; the actual linkage is accomplished by means of tiny hooklets.

The feathers of some orders, the emus for example, are double structures with an aftershaft branching off at the base of the rachis and being of almost equal length with the main feather. This is presumed to be a primitive feature since the aftershaft diminishes as one ascends the evolutionary scale. In the Galliformes the aftershaft is about half the size of the main feather and in many of the advanced Passeriformes it is missing altogether.

This basic pattern has been adapted to produce feathers suited to do various jobs essential to the life of the bird, thus underlining nature's law of always closely linking structure with function.

### THE DOWN FEATHERS

These make up the underplumage of the bird: they are its vital defence against the cold and provide very efficient thermal insulation. Each of these down feathers has a quill and a soft head of fluffy branches, but the barbules are lacking and the feathers are not 'zipped' together as are the larger feathers. This type of feather is particularly numerous in the Anatidae and is often plucked from the breast to line the nest and keep the eggs warm. In the eider *Somateria mollissima* the down feathers have long been gathered commercially to produce eiderdowns and the lining for sleeping bags; the French vernacular name for the species is 'duvet'. Some species have special types of down feathers which break off at the tips, and crush into a fine powder. In many species of hawk powder-down feathers are scattered at random amongst the others whereas in the herons they are arranged in bulk on either side of the breast, flanks and rump. These particular feathers grow continuously to keep up with the rate of fragmentation and soak up water, blood and slime—so much a part of the life of these birds.

### THE CONTOUR FEATHERS

These are the most specialised and have a row of barbs on either side of the shaft producing the vane of the feather (*see* Fig. 52). Each barb is made up of many smaller barbules which, unlike those of the down feathers, actually interlock to produce a smooth surface. Whenever the arrangement becomes disrupted all the bird has to do is to pass the feather through its bill and the barbules become linked once more. The contour feathers vary a lot both in length and thickness and range from the larger and stiffer primary and secondary flight feathers (together called remiges) and the tail feathers (called retrices) to the softer and more delicate feathers which cover the body and give it shape.

THE SEMI-PLUMES

In appearance these are intermediate between contour and down feathers. They are distinguished from down feathers by the rachis being longer than the longest barbs; they occur between the tracts of contour feathers providing fill-in material and efficient insulation, as well as smoothing off the shape of the bird.

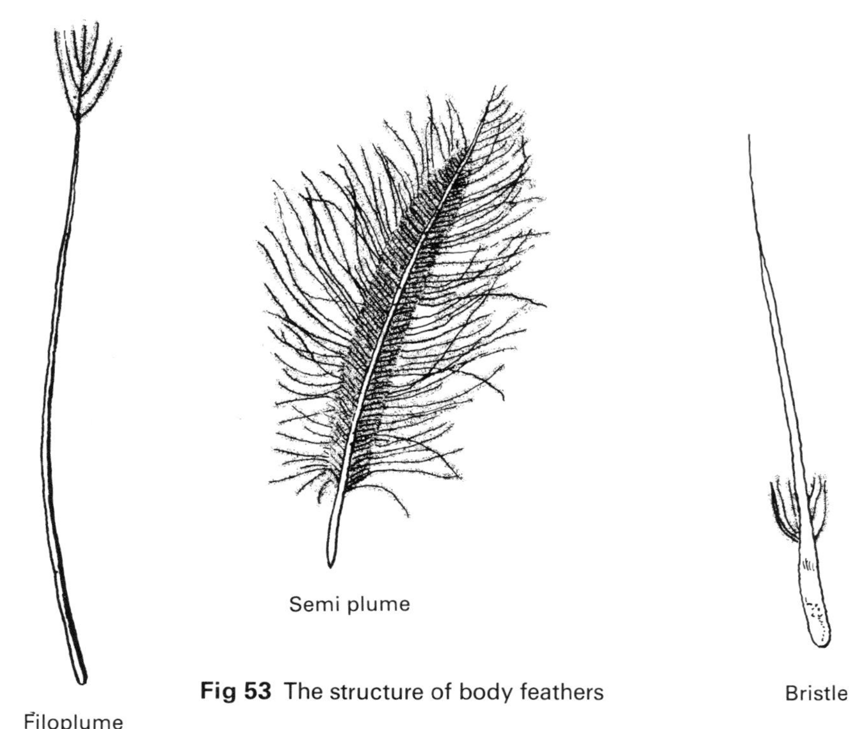


**Fig 53** The structure of body feathers

FILOPLUMES

These are on a very much smaller scale and stand up like hairs, being easily observed if you look at a plucked chicken. They are made up of a thin shaft with a few short barbs or barbules at the tip. It has often been suggested that they are degenerate contour feathers but it now seems that they are sensitive structures assisting in the movement of the other feathers. They have a lot of free nervous endings in the follicle walls, and these are connected to pressure and vibration receptors around the follicle. They may well play a role in keeping the contour feathers in place during preening, display and also in flight.

BRISTLES

These are specialised feathers with a very stiff rachis and few, if any, barbs. They often function as eyelashes and as aids to birds such as nightjars which catch flying insects.

## Arrangement of the Feathers

Only in a few quite primitive orders do feathers grow at random all over the body surface; other species grow their feathers in about eight well-defined tracts called pterylae leaving areas of bare skin between, these being called aptera. Among the aptera are the brood patches, areas of skin rich in blood vessels which enable both the eggs and vulnerable young to be kept warm.

## Numbers of Feathers

There aren't actually 40,000 feathers on a thrush as the old music hall song suggests; the actual number varies from just under 1,000 in some hummingbirds to over 25,000 in some swans during the winter months. In most birds feathers constitute between 15 and 20 per cent of the body weight. Birds have more feathers in winter than in summer which would seem to suggest that insulation as well as flight was a major factor in their evolution.

## Colour

The range of colours in the avian world is very wide and its full significance will be discussed in Chapter 10. The colours are produced by a combination of very few pigments, of two main types. Firstly there are the melanins which are responsible for blacks and yellows, and then there are the carotenoids which also produce yellows as well as some very attractive reds. Nature is always most economical and it is usually only the exposed parts of the feather which are brightly coloured. Melanins are synthesised in the bird's own body, but carotenoids must be obtained either directly or indirectly in the diet, although they may well be subject to physiological changes in the digestive system. Nowhere is the significance of the diet better demonstrated than in the flamingoes which lose their delicate pink colouration when deprived of their natural planktonic diet.

Other colours occur in the avian world in addition to those due to pigments, and are caused by various microscopic structures on the feathers which resemble prisms of wax. These refract light to produce the splendid colours so much a feature of hummingbirds and often of the European kingfisher *Alcedo atthis*. Albinism occurs frequently in birds and is particularly obvious, even though no more common, in dark coloured birds such as the crow *Corvus corone* and the European blackbird *Turdus merula*: albinistic birds may even lack pigment in all feathers, the bill and feet, and even the eyes appear pink due to the haemoglobin in the blood showing through.

Feathers often change colour by abrasion, the ends of the feathers being rubbed off during use; the change from winter to summer plumage is often achieved this way rather than in an energy-sapping moult.

## Feather Maintenance

Birds have to take great care of their feathers if they are to retain their flying efficiency; in heavy rain, for example, some birds can get so wet that they become waterlogged. In many swimming and diving birds the plumage is completely waterproof, and their liberal use of preen oil may confer some advantage. The cormorants, however, have not managed to evolve a completely waterproof plumage and this is why they are often seen standing on a sandbank or perched on a suitable spot literally hanging out their wings to dry. The preen gland (also called the oil or uropygial gland) is the only gland with an external opening so far evolved by birds, but it cannot be, as some writers have suggested, a bird's only defence against waterlogging since many species do not have such a structure. It has been suggested that the oil might be more important as a source of Vitamin D than it is as a waterproofing agent. The powder down also has its part to play in keeping the skin dry, but it is constant attention to the feathers which ensures that one overlaps the other like tiles on a roof, and when all the barbules are 'zipped up' a very efficient waterproofing system is produced.

## Moult

However well they are looked after, feathers become worn with use and age and they need to be replaced, usually once each year. During the moult the old feather either falls out or is forced out by the growth of its replacement. Each new feather starts as a reddish pimple-feather with a grey-coloured sheath, and new material is gradually added from the base; finally the sheath bursts and flakes off, allowing the brand new feather to expand. Once dry the feather receives no further increments of food and if damaged cannot be replaced until the next moult. Should a feather be pulled out, however, it is replaced almost immediately. In the majority of species the moult is a gradual process, only a few feathers being lost at a time, so that the bird is never without adequate insulation and can always use its wings even if they are not at maximum efficiency. In many Anatidae, where there is also an eclipse plumage in drakes, and also in some divers, rails and cranes all the flight feathers are shed at once, but this arrangement is only really satisfactory for water and marshland birds which can find food in relative safety. In the vast majority of species one set of feathers are retained for a whole year, but two exceptions to this rule are the golden eagle *Aquila chrysaetus* which retains some feathers for two years, and *Rhinoflax vigil*, a hornbill which has two central tail feathers but only one of these is moulted each year. Whichever type of moult a bird goes through, the same sequence is followed each moult period.

Birds, even waterfowl, do retain considerable, if restricted, powers of movement during the moult and it is only the penguins which seem to be completely immobilised by it. The Emperor penguin *Aptenodytes forsteri* fasts during the three to five week moulting period during which time it stands on the ice while the newly forming feathers push out the old ones which come away in big patches.

## The Bird's Skeleton

Birds have evolved a skeleton (*see* Figs. 54 and 55) capable of withstanding the strains imposed by the demands of walking and flying—at one moment the whole weight is supported by the hind limbs and the next the wings are propelling and supporting the animal. An obvious adaptation is a shift in the centre of gravity achieved by having a shorter body than either reptiles or mammals whose main mode of locomotion is on all fours. The demands of flight have produced a much more rigid skeleton, but one which is much lighter.

Many bones which are quite separate in other vertebrates are fused in the skeleton of a bird. In the lower portion of the vertebral column the lumbar vertebrae are joined, as are all the bones of the hip girdle thus forming a light but very strong plate which in turn rests on the thigh bones. This arrangement supports the weight of the bird whilst it is on the ground. The vertebrae in the chest region are also fused, and in flying birds the breast bone (sternum) which connects the ribs has a deep keel. This arrangement has two functions; the heart and lungs are protected and the huge flight muscles which power the wings are attached to the keel (*see* Chapter 4). Ratites do not have a keel, but birds such as penguins do because they use their flipper-like limbs to very good effect when swimming, and virtually fly under water.

The ribs of birds also show interesting adaptations to flight. They produce great rigidity by strap-like projections on each rib. These overlap the ribs behind and so form a brace. The breast muscles are huge structures and during flight are subject to tremendous pressures, and so extra support is essential. Thus we have two clavicles (collar bones) which are joined in front of the breast bone to form the wish bone. Birds have yet another pair of bones called the coracoids which also act as braces running upwards from the front of the breast bone and which join together, and with the clavicle and narrow shoulder blade, to form a socket into which the base of the wing fits snugly.

The shoulder blades and collar bones move against the upper ends of the coracoids which join below to form the surface on which extra flight muscles are attached. The feathers are inserted along the length of the 'arm', the secondary feathers being attached to the ulna, and a combination of the bones of two fingers supports the primary flight feathers. The third surviving finger forms the bone of the bastard wing (alula) which acts like an aeroplane's slot during flight and helps to prevent stalling. The study of bird

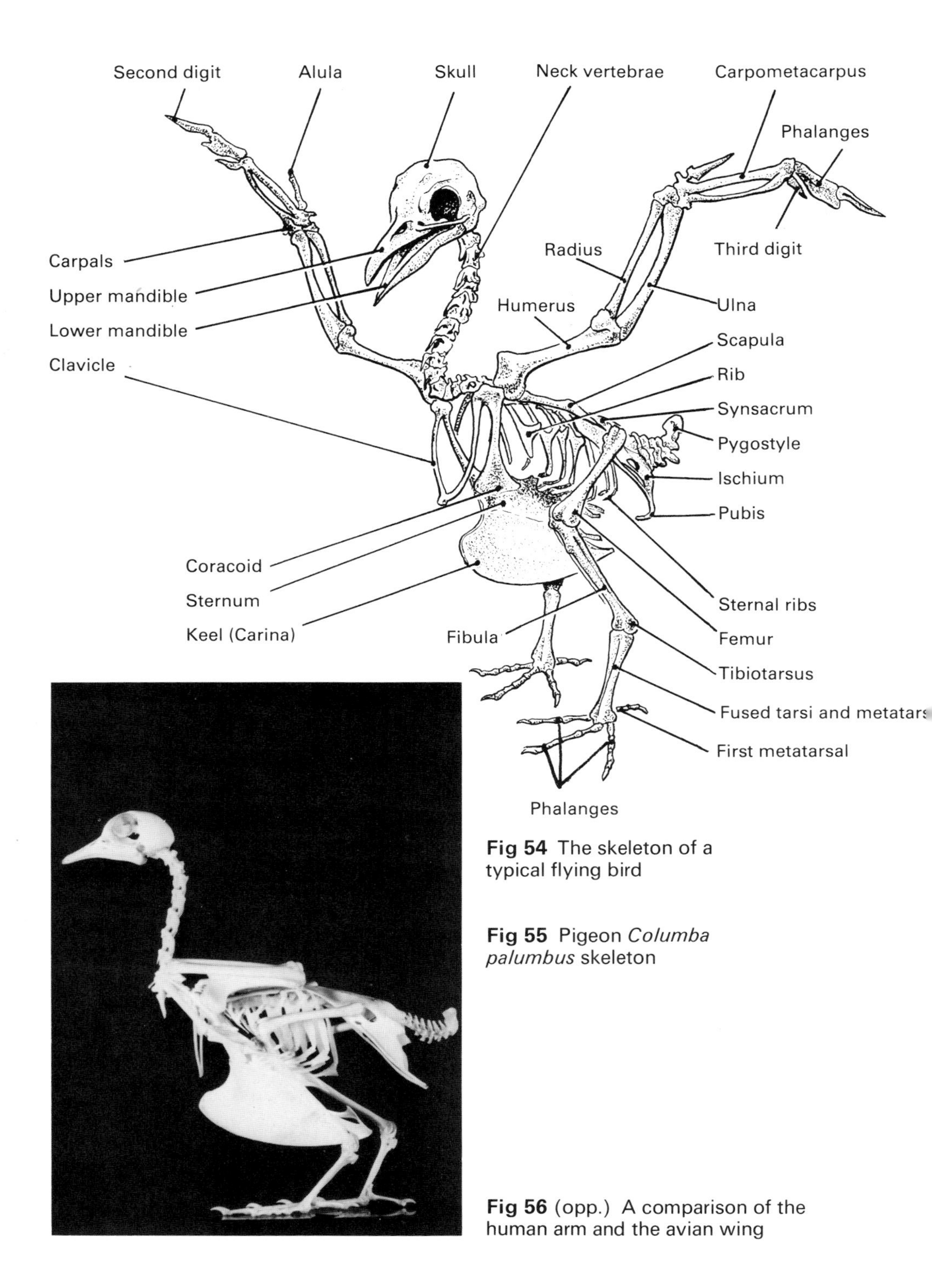


**Fig 54** The skeleton of a typical flying bird

**Fig 55** Pigeon *Columba palumbus* skeleton

**Fig 56** (opp.) A comparison of the human arm and the avian wing

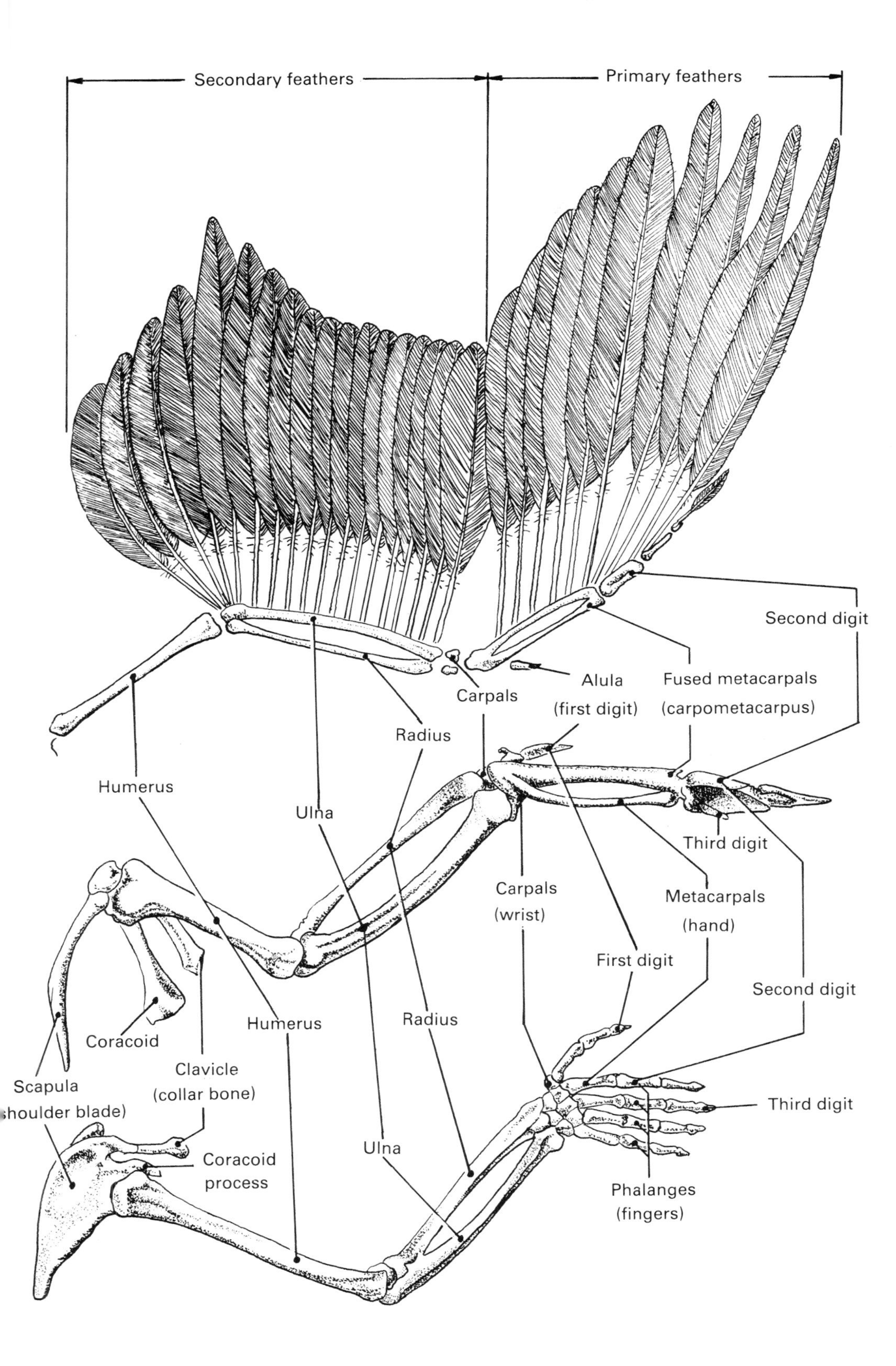

Secondary feathers
Primary feathers
Second digit
Fused metacarpals
(carpometacarpus)
Alula
(first digit)
Carpals
Radius
Humerus
Ulna
Third digit
Carpals
(wrist)
Metacarpals
(hand)
First digit
Second digit
Coracoid
Humerus
Radius
Scapula
(shoulder blade)
Clavicle
(collar bone)
Third digit
Coracoid
process
Ulna
Phalanges
(fingers)

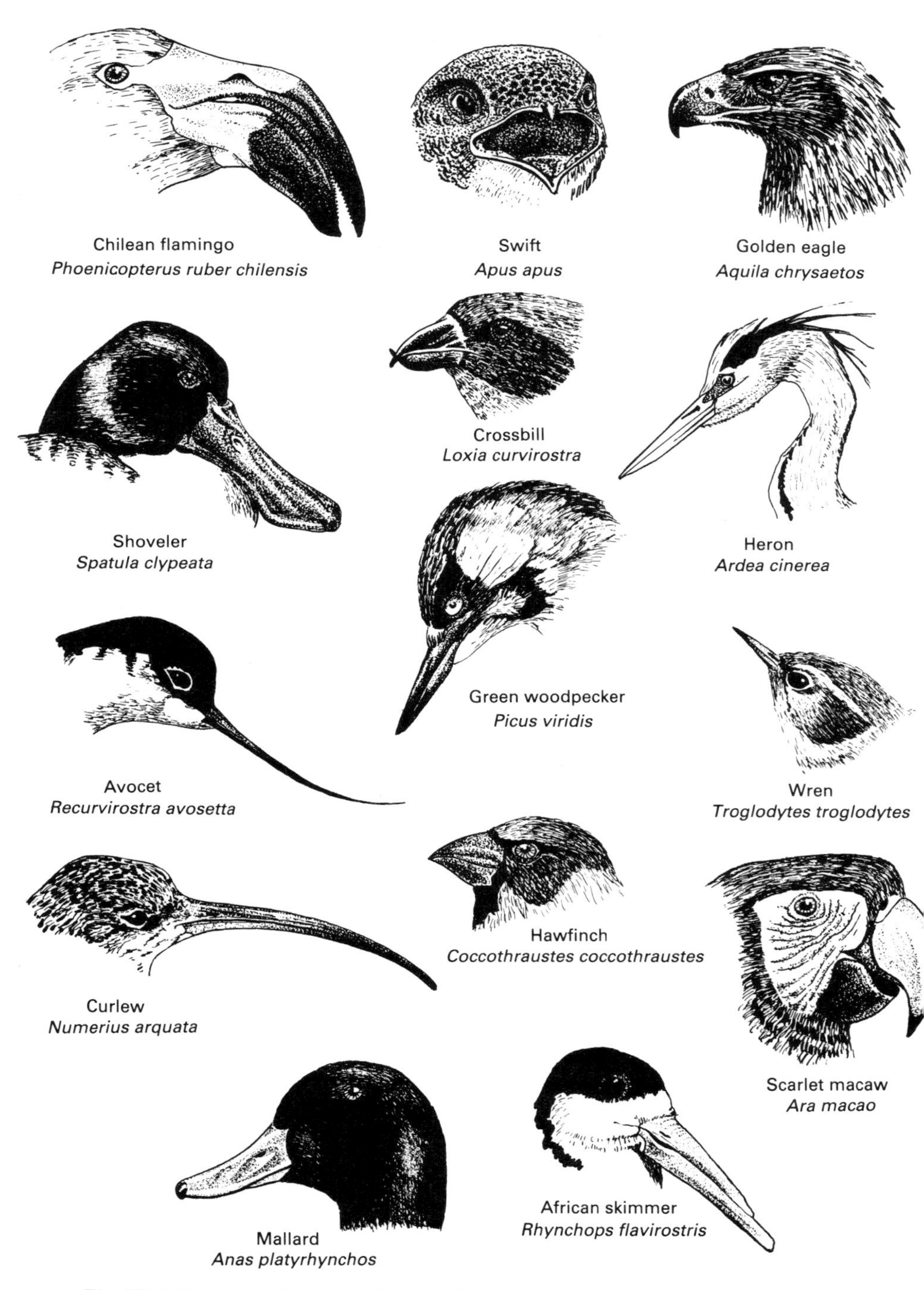


**Fig 57** Bill shapes. Evolution has produced a huge variation in the avian bill, enabling food from many different sources to be exploited

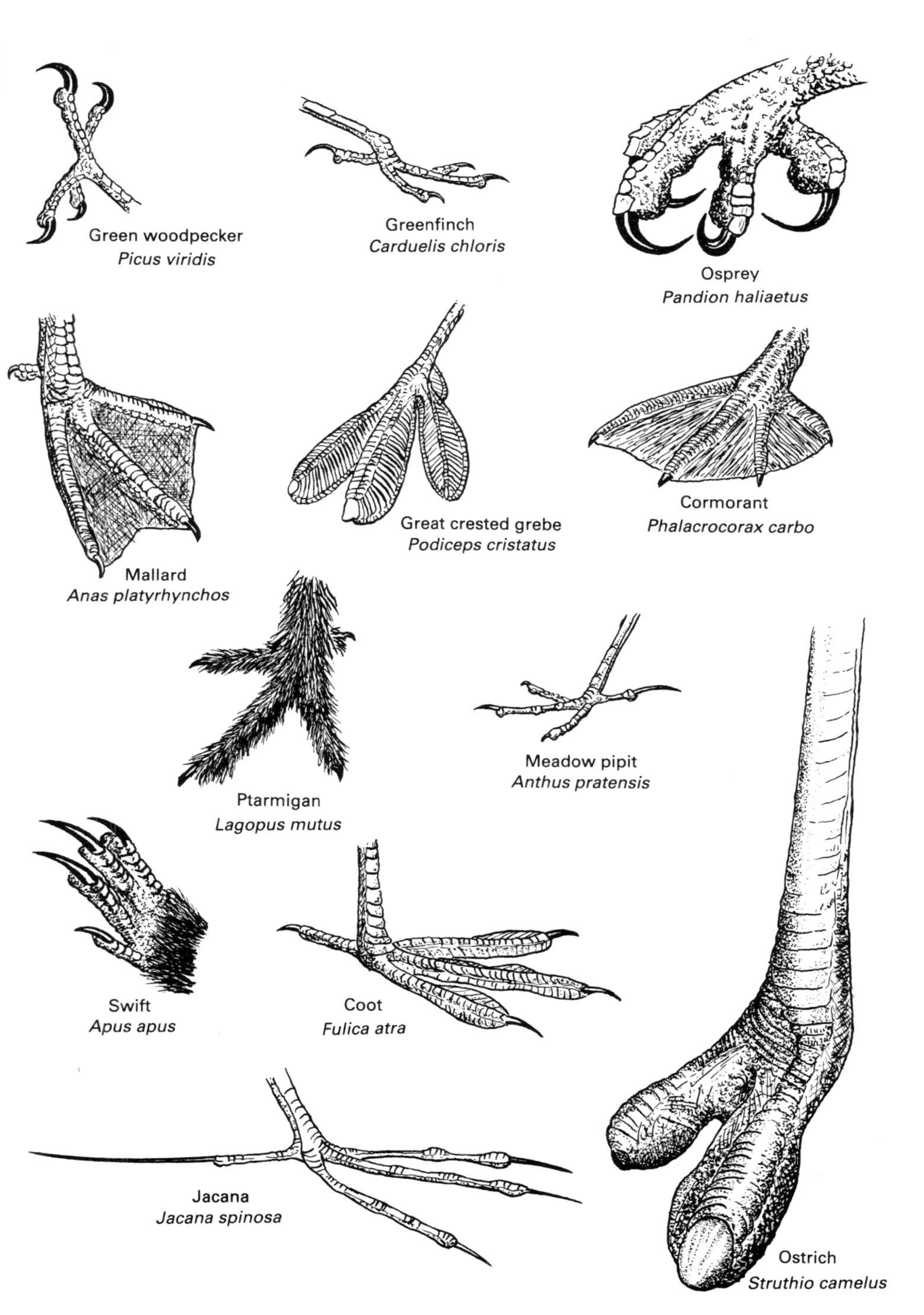


**Fig 58** Foot shapes. The avian foot is required to cope with many different environments and wide variation has evolved

flight, however, demands more attention and forms the subject matter of the next chapter.

The skeleton of the head and neck of the bird has also evolved in response to sacrifices made to achieve locomotory efficiency. The loss of the fore-limbs as manipulatory organs has resulted in the development of long and supple necks enabling the bill to reach all parts of the body. Mammals, even the giraffe, all have only seven neck vertebrae, whereas in birds there is a varying number from 11 to 25. The bill is a horny but very light tool adapted to carry out a remarkable variety of functions and to cope with the particular diet of an individual species (*see* Fig. 57).

It is not only the bills which demonstrate nature's powers of invention, and a look at the feet of the various orders of birds will reveal a similar variation (*see* Fig. 58). Birds may have two, three or four toes, but never five as often seen in reptiles and mammals. The tarsus of a bird's leg (often called the shank) carries out the same job as the human leg, but is, in fact, the anatomical equivalent of our foot. In birds the thigh bone is very short, and is hidden within the body wall (*see* Fig. 54). The ankle region is also greatly simplified but does form a very strong single action joint ideal for withstanding the stresses and strains of take-off and landing. It is in the legs and feet that the reptilian ancestry of the class Aves is most clearly indicated by the presence of scales. Some birds such as the ptarmigan *Lagopus mutus* which live in cold climates cut down heat losses by having feathered feet. The feet of swimming birds are usually webbed in some way or another, in most cases serving to bind three toes together so that they function as a paddle. In the Pelecaniformes all the four toes are connected by webbing, whilst in other orders such as the Podicepediformes (the grebes) and in some Gruiformes (the coots) each toe is itself webbed although not connected with any other. In the case of birds of prey the feet have evolved not in response to locomotion but to obtaining food, and in the owls (Strigiformes) and diurnal birds of prey (Falconiformes) it is the strong talons which kill, the hooked bill being reserved for tearing the flesh.

*Chapter Four*

# Flight

Without doubt the major factor in bird development has been the evolution of flight. It has allowed a more or less free movement over much wider areas than has been possible with purely aquatic or terrestrial organisms; it has also allowed them to occupy the ends of tree branches and keep clear of all other natural predators save other birds. This has meant that they have been able to become noisy and develop a language with which to communicate with others of their species. The need for concealment is also less vital and their often delightful colours account in part at least for the increasing popularity of birdwatching as a leisure occupation.

The technicalities associated with flight have also been solved by bats, but birds have hedged their bets by retaining more use in their hind limbs. Flight, however, has imposed certain limitations, most of which are concerned with weight, but it is also important that the wing should be small enough to fold up neatly, and the associated muscles must be adequate to power flight but not too heavy to hinder take-off. The actual shape of the wing will obviously vary depending upon the bird's mode of life.

Four basic types of wing can be recognised (*see* Fig. 59). Species such as the pheasant, which tend to live in enclosed habitats such as woodland and need to dodge quickly in and out of obstructions, have evolved elliptical wings. These have what ornithologists have called a 'low aspect ratio' which means that length divided by width gives a low number; they are also highly cambered, and the outer primary feathers are slotted. This structure is good for twisting flight but less efficient when it comes to sustained high speed flight. Gamebirds, woodpeckers, cuckoos and many passerines, especially the corvids, tend to have this type of wing.

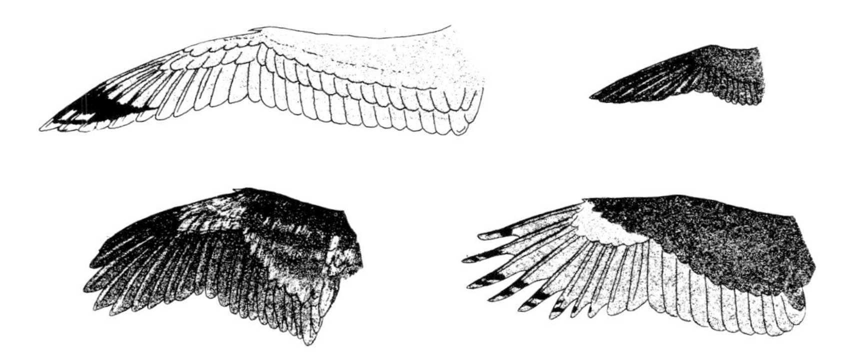

**Fig 59** The four basic wing shapes: *top left*, gliding (e.g. gull); *top right*, high speed flight (e.g. swallow); *bottom left*, rapid take-off (e.g. pheasant); *bottom right*, soaring (e.g. eagle)

Birds of prey, on the other hand, have developed high speed wings (with a high aspect ratio—long and narrow) which have little camber, taper to a point, and tend to be swept backwards as in a high speed jet fighter. All the really fast flying birds have evolved this pattern including swifts, swallows, waders and wildfowl. Seabirds such as the shearwaters and albatrosses also have a high aspect wing but which is long, narrow and flat, being adapted for long distance gliding, and there is no slotting on the primaries. Yet another gliding variation is found in the fourth type of wing which is a slotted high-lifting structure typical of vultures, eagles and storks. Here we find broad wings providing for light loading, but they also need to be short in order to make use of subtle variations in air currents. Each primary feather can be used as a separate aerofoil and thus slotting is a major feature of this type of wing.

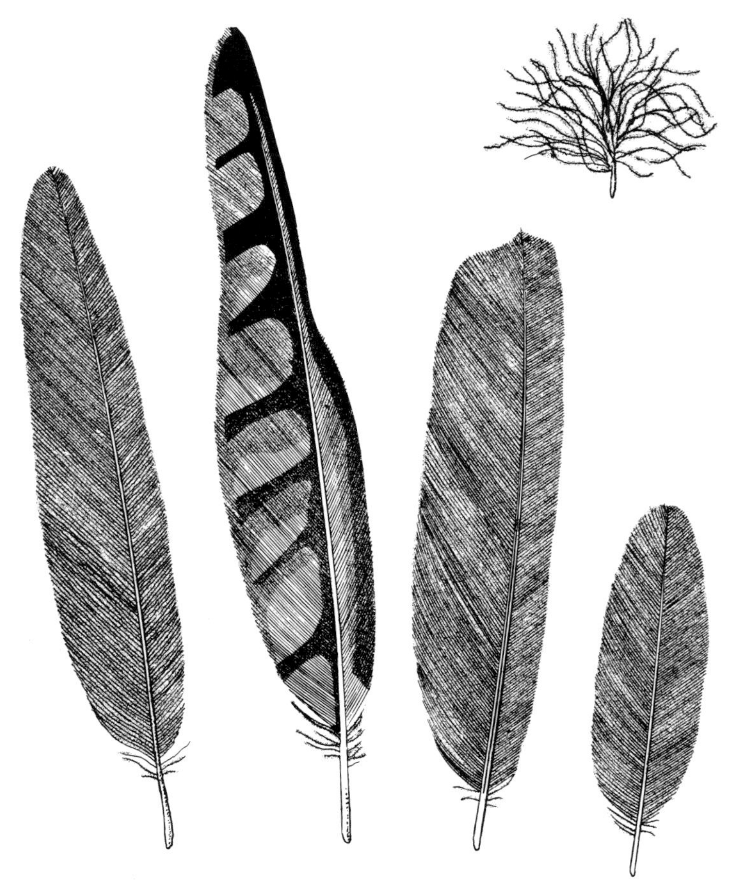

**Fig 60** Types of feather. *Left to right*: primary; emarginated primary of soaring bird; secondary; tail; *and above*, down

As we struggle to understand the precise mechanism of bird flight it is not surprising that we should compare the avian wing with that of an aircraft, but we must take care not to take the analogy too far. An aircraft moves because of the propulsion provided by the engine, and remains in flight due to the aerofoil properties of the solid wings. Birds, on the other hand, have wings of feathers which are responsible for both the propulsion and the efficient aerodynamics. The primary feathers are attached to the hard bones, and these are driven through the air by large flight muscles attached to the keel. If these flight feathers are damaged or lost then the bird will find flight impossible. The secondary feathers are inserted along the arm (*see* Fig. 56) and are responsible for lift; it has been experimentally proved that if half these feathers are removed the bird is still able to fly, but some control will be lost. Each feather functions independently and allows the shape of the wing to be altered during flight allowing the bird much freedom of movement—a facility denied to even the most sophisticated of our aircraft with their solid aerofoils. It is not just the feathers of the wing which play a vital role in flight—the tail feathers (retrices) are used as a rudder allowing twisting and turning. They also act as an efficient brake prior to landing. Although there are many differences between the aerofoil of an aircraft and a bird's wing a study of aircraft design can throw a great deal of light on avian flight and *vice versa*. It will, however, explain only gliding flight which is relatively passive; the much more active flapping flight has never been successfully copied by human technology and is therefore more difficult to explain satisfactorily, as indeed is hovering, the third type of flight.

## Gliding Flight

A simple wing is merely an extension of the body, and if this is held edge on to a current of air, the force of the current will tend to drag the wing with it downstream. Only some external force pulling the wing upstream can hold it in position. If the wing is raised, the edge facing upstream will tend to be lifted, although at the same time the force tending to drag it downstream increases. Thus we have two forces which are called 'lift and drag' (*see* Fig. 61).

When a wing is held at a slight angle to an air current the air flows faster over the upper surface than it does over the lower (*see* Fig. 62), thus creating a loss in pressure above the wing causing 'lift'. At the same time resistance to the moving air tends to drag the wing backwards. The combined effect of these two forces is to lift the wing and drag it backwards. True gliding flight is only possible when lift and drag forces are so adjusted as to be equal to the weight of the bird. The object of the development of any bird's wing—and evolutionary pressures are obviously working to this end—is to increase lift and minimise drag. Let us now see how these facts can account for the gliding flight of a bird such as a gull.

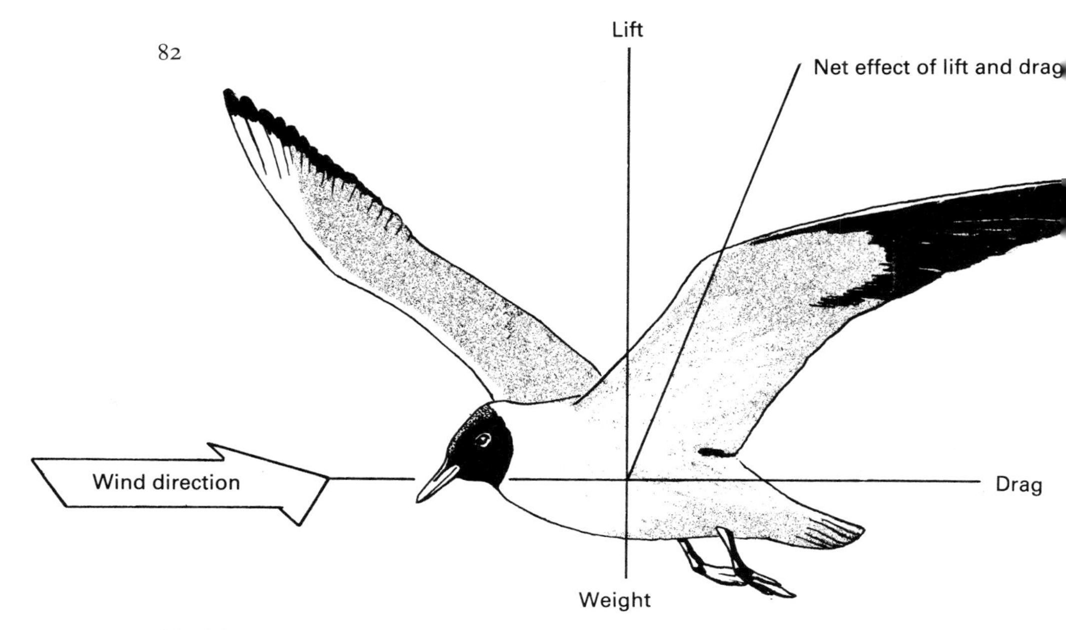


**Fig 61** The forces of lift and drag affecting normal flight

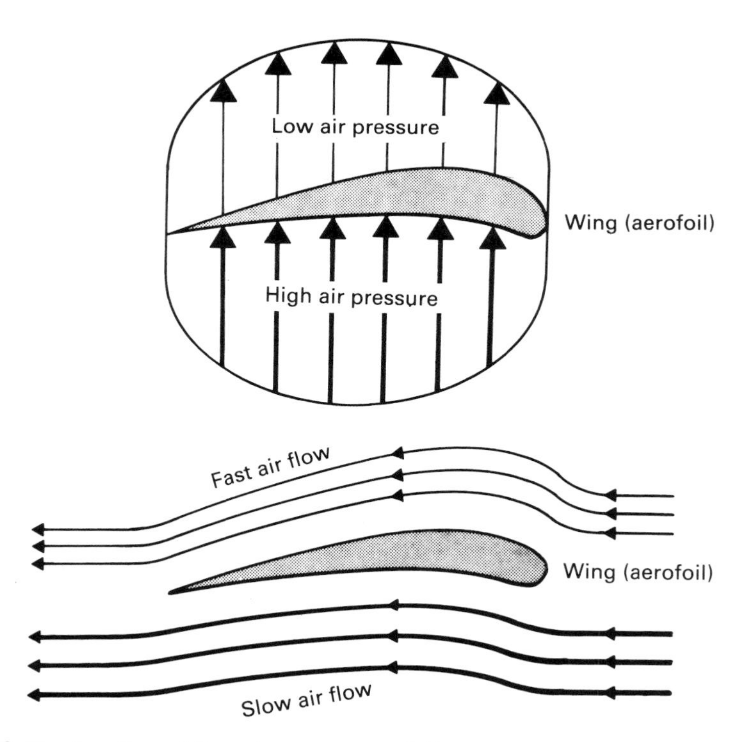


**Fig 62** The function of the wing as an aerofoil

The wings are stretched out stiffly, but the bird is gradually losing height due to gravitational pull; gravity, however, bestows the advantage of acceleration. A good glider travels a long way horizontally with minimal loss of height and it is possible to measure the efficiency of a glider by calculating the angle between the track of its motion and the level horizon (*see* Fig. 63). This angle does not depend upon the weight of the bird, but depends totally upon the forces of lift and drag, although the shape of the wings, as outlined in Figure 59, will play a part in this.

The speed of the glide on the other hand does depend to some degree on the weight of the bird—a heavy bird with small wings will glide quickly, whilst a light bird with large wings will glide very much more slowly. The actual distance travelled must depend to some degree on the height from which the bird starts. This was obviously the first problem to be solved by ancestral birds such as Archaeopteryx as well as by other flying creatures without the advantage of powered flapping flight.

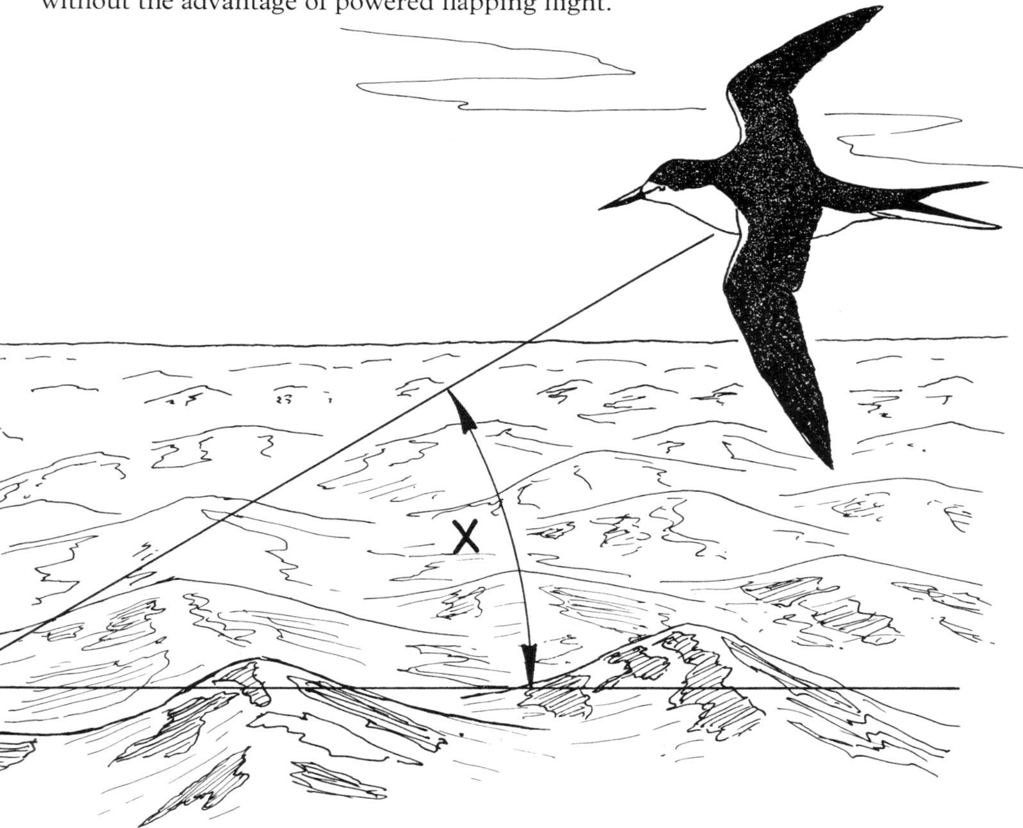


**Fig 63** The bird as a glider. The lower the value of X, the more efficient the glide, and the less energy the bird will consume

Sooner or later, in still air, the bird must reach the earth and the glide is at an end *but* if the air is not still and is itself moving relative to the ground then the story is very different. If a bird is gliding downwards, losing height at say 5 metres per second and it meets an air current rising from the ground also at 5 metres per second then the bird will glide level. Should it meet a rising current of air in excess of 5 metres per second then the bird will rise without expending any of its own energy. There are two main causes of upward air currents and both are eagerly exploited by birds.

Obstruction currents (*see* Fig. 64) are caused when a wind drives against a large solid object and is pushed upwards. They are used by swifts *Apus apus* and kestrels *Falco tinnunculus* which glide along the windward side of the eaves of buildings, by seabirds soaring above a line of cliffs when an onshore wind is blowing and also by birds of prey and ravens *Corvus corax* literally hanging around mountains.

Thermal currents are caused when air, warmed by the hot surface of the earth or even as a result of human activities, moves upwards and is replaced by colder air dropping from higher levels in the atmosphere (*see* Fig. 65). In temperate climates the use of thermal up-currents by gliding birds is somewhat restricted, but nevertheless birds may sometimes gain considerable heights over towns and other regions where hot air is rising. In tropical countries vultures with their broad wings make good use of quickly rising thermals which are generated surprisingly quickly as the hot rays of the

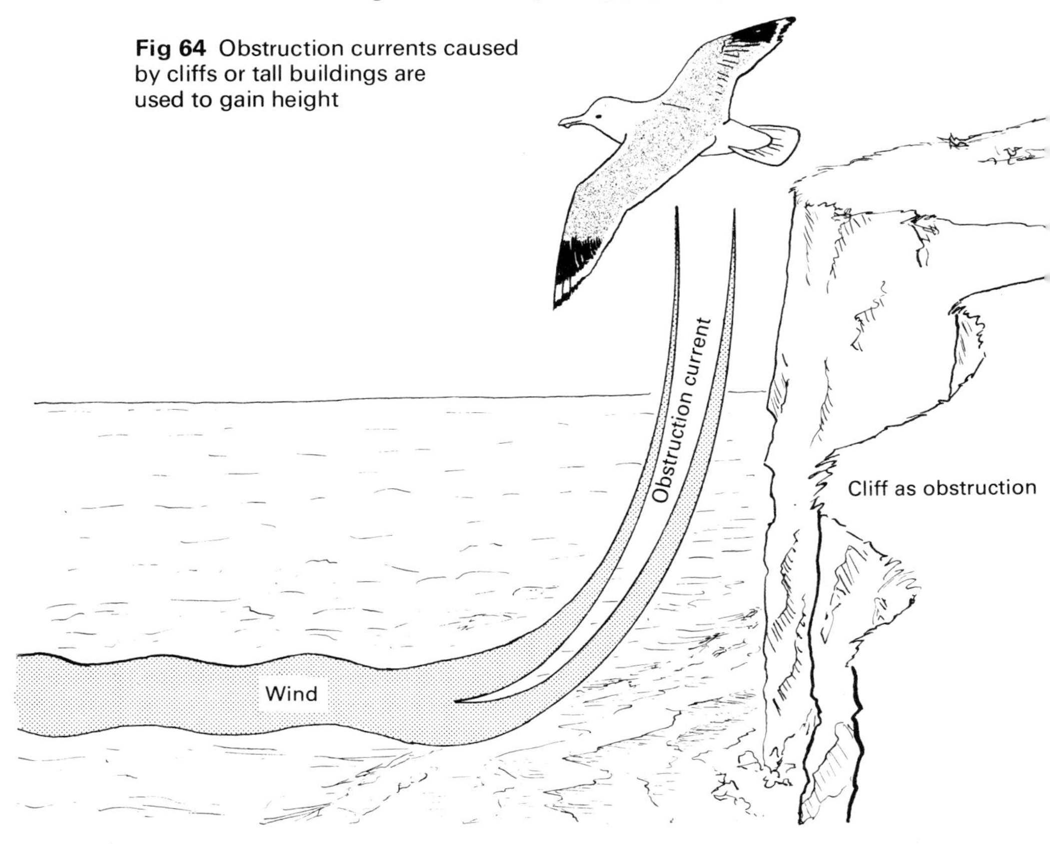


**Fig 64** Obstruction currents caused by cliffs or tall buildings are used to gain height

morning sun strike the cooler earth. The birds may then pick up a belt of colder air and spiral downwards with it before picking up yet another thermal and travelling aloft once more.

This description could well have been taken from a glider pilot's manual, but certain seabirds, particularly the albatrosses and shearwaters, have developed a type of glide which does not depend upon the presence of either thermal or obstruction currents. The precise explanation of this type of glide has proved very elusive but it has been called 'gust flying'—rising into the gust and falling on the lull. This idea is supported by the knowledge that the albatross cannot fly in the absence of wind, a factor of some importance with regard to its distribution. Its flight follows a fixed pattern which involves turning into the wind at the end of a rapid descent downwind. Its speed relative to the air has been increased sufficiently to build up its kinetic energy (the energy of movement) and this can then be used to drive the bird upwards. This in turn increases the bird's potential energy (the energy due to its position; the higher it is the more energy it possesses). This potential energy can then be used to power another slanting dive and the process can be repeated just so long as there is sufficient difference in wind speed between sea level and a height of, say, 20 metres above it.

We are able, then, to go quite a long way along the road towards understanding gliding flight, but things become much more complicated when we come to a consideration of flapping flight.

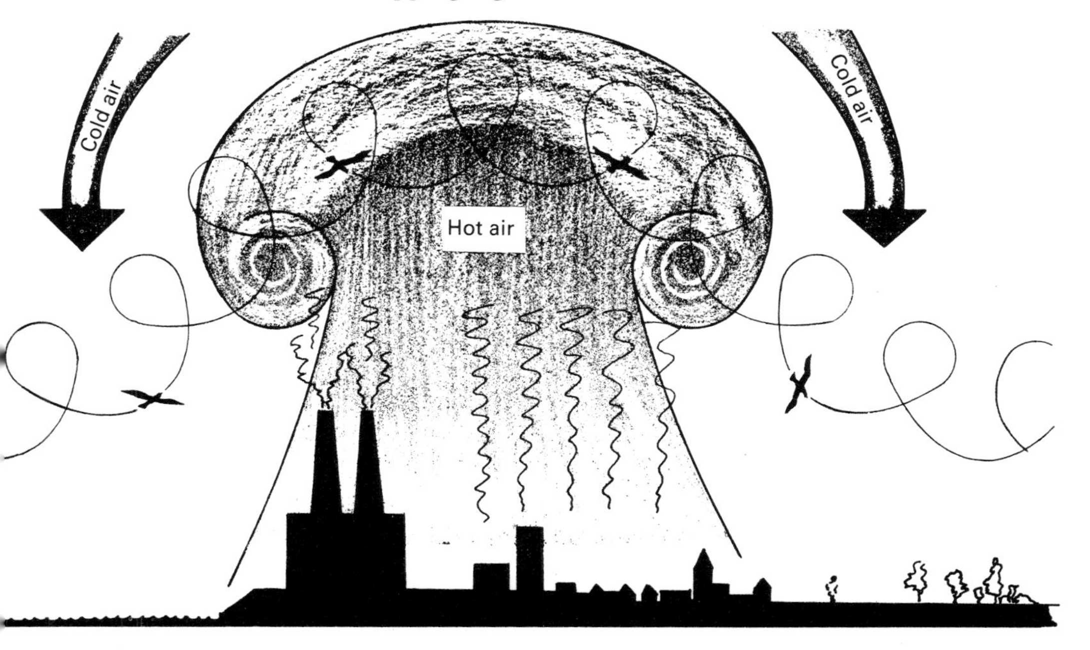


**Fig 65** Thermal currents caused by man or nature can be used to gain height

## Flapping Flight

Two distinct movements must be considered here: the power stroke moves forwards and down, and the backstroke has the function of returning the wing to the position from which the next power stroke commences. During the glide the inner wing and the 'hand region' are spread to produce a continuous aerofoil. During a flap, however, the two parts carry out different functions; the inner wing (secondary feathers) gives lift whilst the hand section (primary feathers) pulls the body forwards. During the power stroke the primary feathers are linked together to produce a near perfect aerofoil giving maximum thrust and minimum drag. In the smaller birds the primary feathers are separated on the upstroke, functioning like the slots in a Venetian blind and allowing air to pass through and thus considerably reducing drag (*see* Fig. 66). This mechanism has not proved suitable either for the larger birds such as the gulls or for the small but long-winged birds of the swallow tribe; these species either flex or partially close their wings on the upstroke. The modern technique of slow motion movie photography has added, and is still adding, significantly to our knowledge of bird flight.

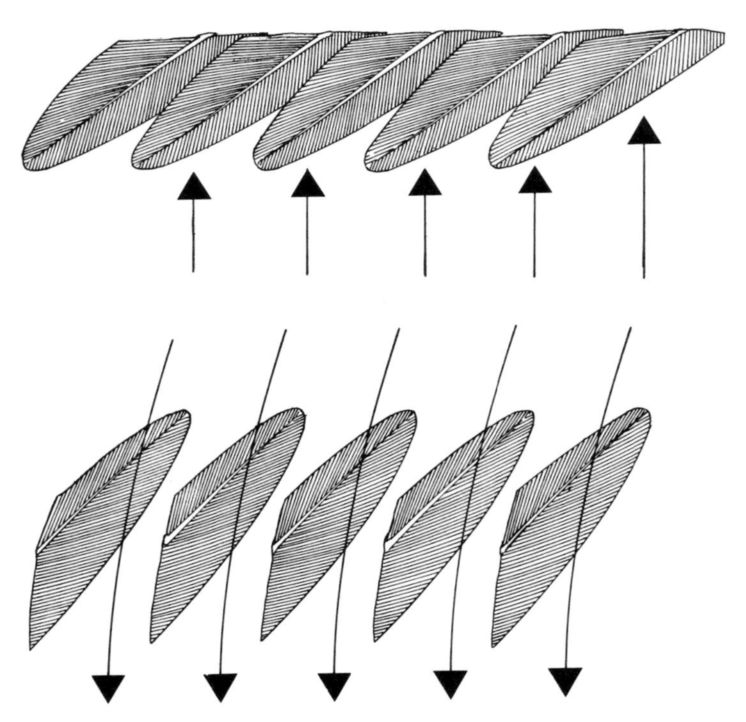

**Fig 66** The relative positions of the primary, or flight, feathers on the downstroke and the upstroke during flapping flight

Despite the fact that all species employ similar basic techniques in flapping flight, some birds are much more efficient than others and we must look for some structural feature to account for this. We find it in what is known scientifically as wing loading. A bird with a large wing and a light body will obviously fly better than one with a heavy body and a wing with a small surface area. We can express flying potential in the form of an equation by dividing the wing area by the body weight—this is the wing loading.

$$\frac{\text{Wing Area (in cm}^2\text{)}}{\text{Weight (in gm)}} = \text{Wing Loading (cm}^2\text{/gm)}$$

Table 4 shows wing loading values of various species. From this it can easily be seen that birds which are aquatic or terrestrial feeders have lower wing loadings than those which feed on the wing. A comparison of the swallow's 7.3 and the starling's 3.9, and the great northern diver's 0.6 illustrates this very clearly.

It is not only the wing loading that is important for efficient flying but so too will be the number of wing beats. Two species of similar size might occupy different ranges by one having a faster wing beat than the other and thus increasing its migratory efficiency. The wing beats per second for a

Table 4
SHOWING COMPARATIVE WING-LOADINGS IN SELECTED SPECIES

| Species | Weight (gm) | Wing Area ($cm^2$) | Wing Loading $cm^2/gm$ |
|---|---|---|---|
| *Gavia immer*–great northern diver | 2,425 | 1,358 | 0.6 |
| *Diomedea exultans*–wandering albatross | 8,502 | 6,206 | 0.7 |
| *Hydrobates pelagicus*–storm petrel | 17.4 | 100 | 5.7 |
| *Phalacrocorax carbo*–cormorant | 2,115 | 1,967 | 0.9 |
| *Ardea cinerea*–heron | 1,408 | 3,590 | 2.5 |
| *Anas platyrhyncos*–mallard | 1,408 | 1,029 | 0.7 |
| *Buteo buteo*–buzzard | 1,072 | 2,691 | 2.5 |
| *Perdix perdix*–grey partridge | 387 | 433 | 1.1 |
| *Gallinago gallinago*–snipe | 95.5 | 244 | 2.5 |
| *Larus argentatus*–herring gull | 850 | 2,006 | 2.4 |
| *Tyto alba*–barn owl | 279 | 1,163 | 4.2 |
| *Apus apus*–swift | 36.2 | 165 | 4.5 |
| *Chaetura pelagica*–chimney swift | 17.3 | 104 | 6.0 |
| *Dendrocopos major*–great spotted woodpecker | 73 | 238 | 3.3 |
| *Hirundo rustica*–swallow | 18.35 | 135 | 7.3 |
| *Corvus corone*–carrion crow | 470 | 1,058 | 2.2 |
| *Pica pica*–magpie | 214 | 640 | 3.0 |
| *Parus major*–great tit | 21.45 | 102 | 4.7 |
| *Turdus merula*–blackbird | 91.5 | 260 | 2.8 |
| *Sylvia communis*–whitethroat | 18.65 | 87.1 | 4.7 |
| *Sturnus vulgaris*–starling | 84 | 190.3 | 3.9 |
| *Fringilla coelebs*–chaffinch | 21.15 | 102 | 4.8 |

*After Magnan (1922) Poole (1938)* see *Bibliography*

number of species are given in Table 5.

The whole object of flapping flight is to move the bird from place to place with the greatest degree of efficiency, and a great deal of attention has been focused on the calculation of the speed of flight, but this is by no means easy. All sorts of factors affect it—wind speed, the bird's own state of health, presence of predators or rivals, and whether on migration or not. As a general rule birds on migration fly faster than those not so engaged. Assuming calm weather and level flight Table 6 would seem to be a fair estimate.

Table 5
SHOWING VARIATION IN WING BEAT FREQUENCY IN SELECTED SPECIES

| Species | Wing Beats per second |
|---|---|
| *Podiceps cristatus*–great crested grebe | 6.3 |
| *Phalacrocorax carbo*–cormorant | 3.9 |
| *Phasianus colchicus*–pheasant | 9.0 |
| *Fulica atra*–coot | 5.8 |
| *Falco peregrinus*–Peregrine | 4.3 |
| *Charadrius hiaticula*–ringed plover | 5.3 |
| *Fratercula arctica*–puffin | 5.7 |
| *Larus argentatus*–herring gull | 2.8 |
| *Columba palumbus*–wood pigeon | 4.0 |
| *Turdus merula*–blackbird | 5.6 |
| *Sturnus vulgaris*–starling | 5.1 |
| *Pica pica*–magpie | 3.0 |
| *Corvus frugilegus*–rook | 2.3 |
| –hummingbird, various spp. hovering | 22–79* |
| *Selasphorus rufus*–rufous hummingbird | 200* |
| *Parus caeruleus*–blue tit | 25** |

** from Alexander Skutch, 1974, The Life of the Hummingbird*
*** Personal recording slow motion photography*
*(from Blake, 1947, Auk 64: 619–20; Meinertzhagen, 1955, Ibis 97: 11–114)*

Table 6
SHOWING COMPARATIVE SPEEDS OF FLIGHT OF SELECTED SPECIES

| 15–30 kph (10–20 mph) | 30–50 kph (20–30 mph) | 50–65 kph (30–40 mph) | 65–100 kph 40–60 mph) | 100 kph (60 mph)+ |
|---|---|---|---|---|
| house sparrow | American robin | starling | falcons | swifts |
| wren | heron | mourning dove | ducks | |
| spotted fly catcher | pelican | chimney swift | geese | |
| | | | rockdove | |

At certain times hungry predators can turn on a most impressive burst of speed and the peregrine falcon *Falco peregrinus* whilst stooping at an angle of about 45° is thought to approach speeds of 320 kph (200 mph) and according to some authorities the swift *Apus apus* may exceed even this impressive figure. This fastest of flapping fliers is a member of the Apodiformes some of which have also evolved the most complex of all aerodynamic techniques—the art of hovering flight.

## Hovering Flight

Although they are obviously heavier than air the hummingbirds often seem in complete equilibrium with it. A hummingbird can remain stationary in the air and it can move backwards or forwards, right or left, up or down, and can even turn upside down—a unique achievement. Two equally astounding aspects regarding the flight are how quickly it can stop, and also how quickly it can accelerate to reach maximum speed from the moment of take-off.

In order to understand just how all this is possible we must consider the structure of the wings and the muscles which drive them. The first skeletal

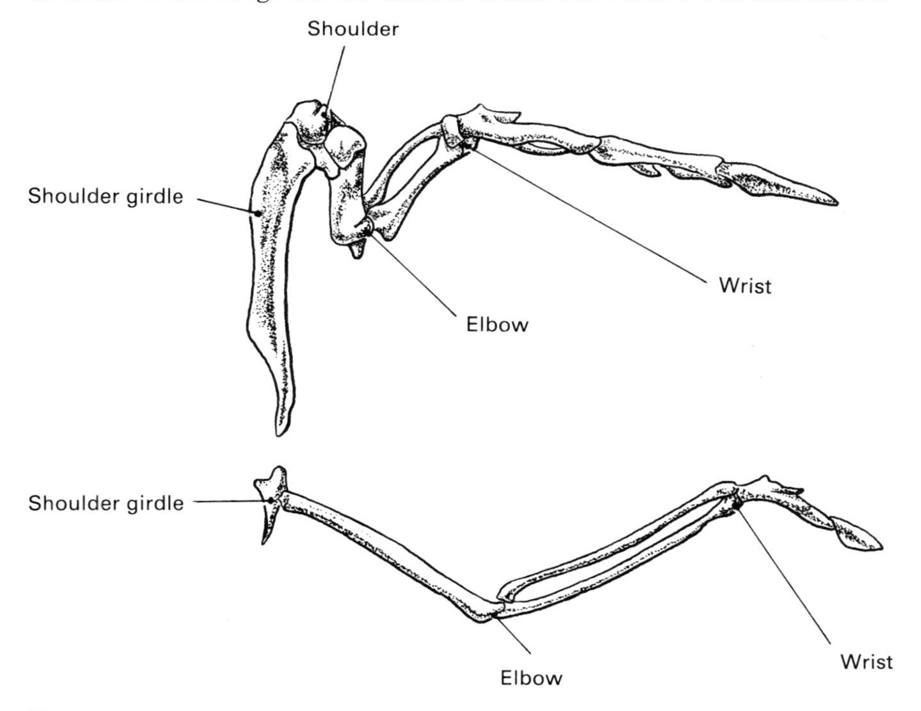


**Fig 67** The variation in wing bone anatomy throughout the bird world is emphasised by a comparison of these two very different wing structures: a pelican above and a hummingbird below

adaptation which becomes apparent is the extreme development of the area corresponding to the hands, whilst the arm area is greatly reduced when compared even to other birds (*see* Fig. 67). This structural change is reflected in the number of feathers: attached to the 'hand' of the hummingbird are ten primaries, the same number as in the soaring albatross, but the real difference is seen in the secondary feathers attached to the 'arm' where the hummingbird's six or seven is far outnumbered by those of the albatross which has about 40.

The short arm bones are arranged in the shape of a very firm V, which is in total contrast to the shoulder joint which is so supple that it will not only allow movement in all directions but will also accommodate axial rotation through 180°. All this means that the tips of the wings are capable of achieving a great deal of controlled movement—absolutely vital to the life of nectar-feeding hummingbirds.

Let us now turn our attention to the musculature of the family and this is equally impressive. The breast bone is large, and the keel is deep. The mighty flight muscles which make up 30 per cent of the body weight are attached to this keel. Birds have two sets of muscles operating the wings (*see* Fig. 68), the depressors which power the downstroke, and the elevators which lift the wing. In most birds the elevators are between five and ten per cent of the weight of the depressors, but in hummingbirds this figure rises to almost fifty per cent and this is one of the main reasons for the

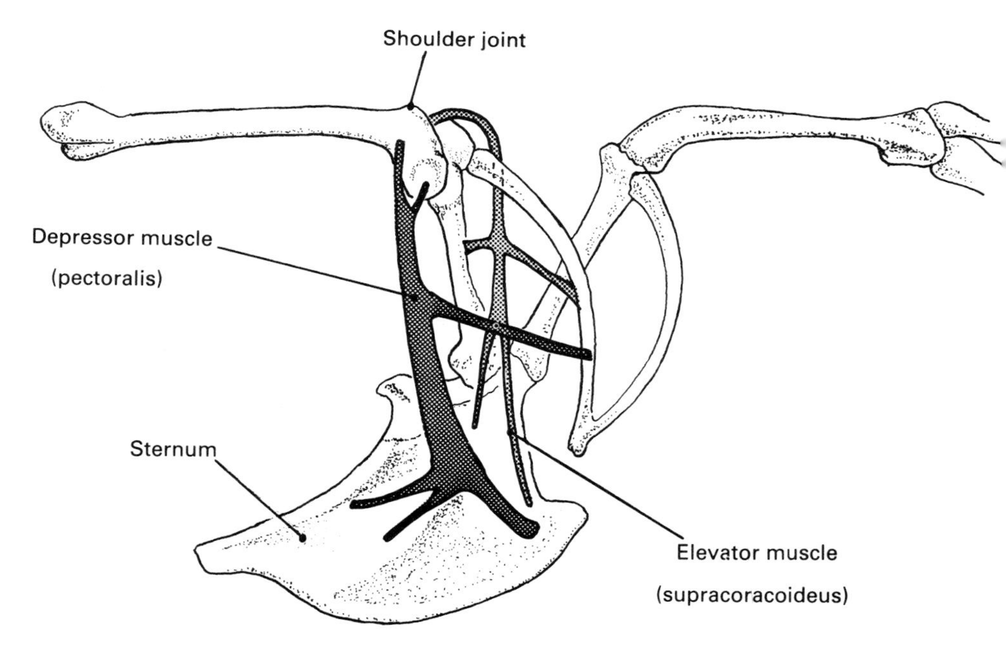


**Fig 68** The flight muscles shown on the upstroke of flight

hummingbirds' exceptional control over their flight. Normally only the downstroke provides lift, the backstroke being one of recovery in preparation for the next downstroke. In the hummingbird the angles through which the wing can be twisted and rotated, by means of the big elevators, can convert even the upstroke into a power movement providing both lift and propulsion.

Thus the hummingbird is able to hover in perfectly still air, its quivering wings moving rapidly backwards and forwards rather than up and down, the tips of the primary feathers tracing a figure of eight. Every time the beat is reversed the wings are pivoted through 180°; this ensures that the front edge always leads, and on the backstroke it is always the undersides of the flight feathers which are on top (*see* Fig. 69). This means that although forward and backstrokes both produce lift the two actually cancel each other out and leave the crafty bird still on station with no movement great enough to disturb its long nectar-seeking tongue thrust deep into its chosen plant.

It should be mentioned that other types of flight are often referred to as 'hovering'. The kestrel *Falco tinnunculus* and at times the kingfisher *Alcedo atthis* are adept at this, but it does not constitute true hovering. Hummingbirds are able to maintain their hover in perfectly still air, a feat totally beyond the kestrel which true to its vernacular name of windhover cannot function unless a breeze is blowing, even though this can be so slight that at ground level it may not be noticeable to a human observer.

We have seen that birds have solved all the problems associated with powered flight. It is very apparent that such an energy-demanding mode of locomotion will require two things—a highly efficient system for obtaining oxygen, and a digestive system capable of speedily processing large amounts of food. These systems are the subject of the next two chapters.

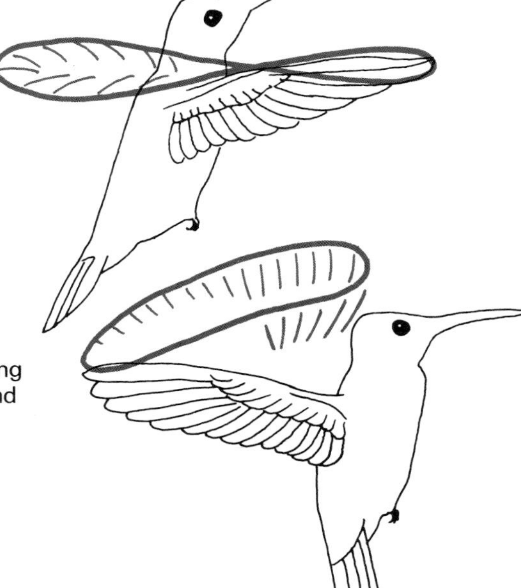

**Fig 69** The wings of the hummingbird are pivoted through a figure of eight, thus allowing the downstroke to cancel out the upstroke and leave the bird effectively motionless, even in still air

*Chapter Five*

# Breathing, Respiration and Blood Supply

Let us firstly avoid the confusion which can arise between the terms respiration and breathing. Each living cell in the higher organisms can be considered as a continuously functioning, self repairing machine. To keep going it requires a supply of fuel (food) and this is 'burned' in oxygen to produce the energy required to power the cell. This is respiration and it can be regarded as a chemical reaction, represented very simply by the equation:

Glucose + oxygen → carbon dioxide + water + energy
(food)

Breathing is the method whereby the respiratory gases are brought into and pushed out of the organism; this process is therefore purely physical and the word breathing is slowly being superseded by ventilation, which is much more explicit.

## Ventilation in Birds

Because we are mammals we tend to assume that the possession of a pair of lungs functioning like bellows pumped by the diaphragm muscle is the only suitable method of ventilation. Let us consider the case of a bird, which weight for weight must require more energy when in flight than any terrestrial mammal. There is also another, much more significant point: birds, especially when migrating, fly at altitudes where oxygen is in such short supply that no mammal could possibly survive. There must therefore be a fundamentally different system in operation, a fact which has been known for over 200 years. In 1758 John Hunter performed what seems to us to be a most cruel experiment in order to show that a bird with a blocked windpipe could still breathe providing there was a connection between one of its bones and the outside air. Hunter wrote

> I next cut the wing through the Os humeri [the main wing bone] in another fowl and tying up the trachea . . . found that air passed to and from the lungs by the canal in this bone. The same experiment was made with the Os fermoris [the main leg bone] of a young hawk and was attended with a similar result.

Thus instead of finding marrow in these bones we find air, and this is true of most bones including the ones in the skull (*see* Fig. 70). As a general rule the more efficient the flier, the more hollow bones it seems to possess. This will tend to give two advantages: the bones will be lighter and make flying more efficient, but also extra space will be available to carry an extra oxygen load.

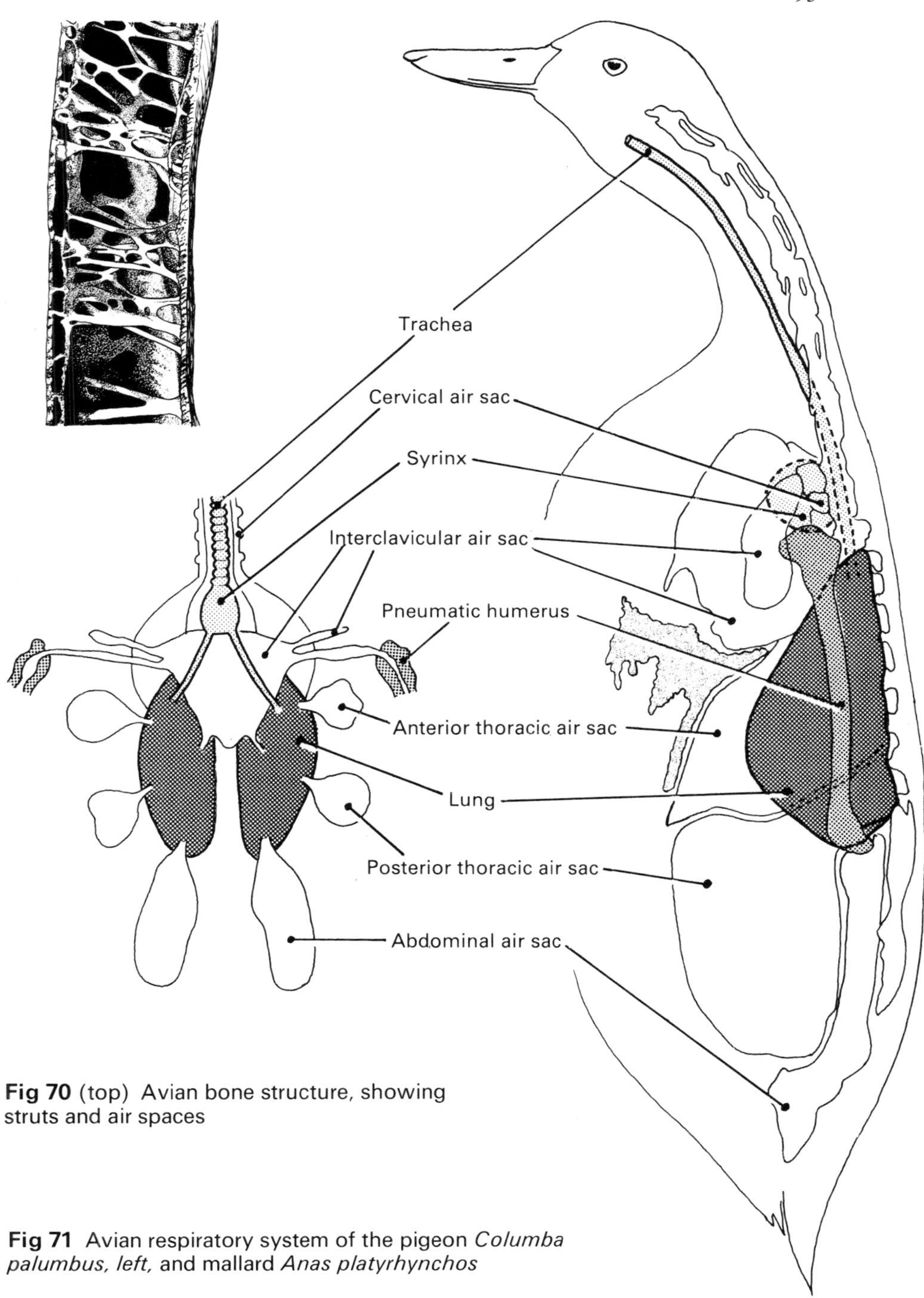


**Fig 70** (top) Avian bone structure, showing struts and air spaces

**Fig 71** Avian respiratory system of the pigeon *Columba palumbus, left,* and mallard *Anas platyrhynchos*

This is not the end of the story, however, and a great anatomist working a century before Hunter came up with another fascinating discovery. William Harvey—the one who unravelled the problems associated with the understanding of the circulation of our own blood—dissected many birds in his thirst for knowledge. He noticed that birds, like mammals, have two lungs connected to the exterior by the trachea (windpipe) but here the similarity ends. In a bird the lungs are connected, at the end opposite to the trachea, to several thin-walled air sacs which fill a large proportion of the chest and abdominal cavities. These sacs are also connected to the air spaces in the bones. Thus the early work of Harvey and Hunter has given us a picture of a complex ventilating system *not* involving a diaphragm but having in addition to lungs, air sacs and pneumatised bones (*see* Fig. 71).

### The Path taken by Oxygen from the Atmosphere to the Bird's Blood

Air with its 21 per cent oxygen enters the trachea through a pair of nostrils situated at the base of the upper mandible, except in the kiwi where the openings are placed at the tip of the bill. In the gannet *Sula bassana* the external nostrils are closed and their function is assumed by secondary nostrils which are situated in the gape. Even these are provided with some protection in the form of a lid which closes when the gannet plunge-dives

**Fig 72** Gannet *Sula bassana*

and probably remains closed during the time the bird is under water. A similar arrangement is found in some cormorants, obviously in response to the same problem.

Eventually the trachea divides into two tubes which are called primary bronchi (or mesobronchi). These lead into the abdominal air sacs, which are easily the largest of the bird's air sacs, and they are fed from the primary bronchus as well as by a number of secondary bronchi (or ventrobronchi) which spread over the lower surface of the lung. Some of these continue onwards to supply the anterior air sacs. Also leading from the primary bronchus are between seven and ten dorsobronchi which spread over the dorso-lateral (back and side) lung surface. The final connection in the network is achieved by tertiary bronchi (or parabronchi) which serve to connect the ventrobronchi with the dorso-lateral bronchi. The parabronchi are, however, important structures in their own right because arising directly from them are structures called air capillaries which are responsible for the gas exchange between the ventilation system and the blood which will transport the oxygen to the body cells where it is needed for respiration.

The avian lung has the very strange characteristic of allowing air to pass directly through it. In contrast, the mammalian lung is like a hollow sac into which push tubes, and these end in structures called alveoli into and out of which air can flow, though it can never flow through them. The bird lung is actually punctured by the very fine branches of the parabronchi, and some workers have likened the passage of air to water moving through a sponge. This has led to the idea that the air sacs might very well function like bellows to drive the air through the lungs.

Things in nature are not always what they seem, and before accepting this theory it is necessary to show that the air sacs themselves do not function as organs of respiratory exchange—as extra lungs in fact. Their walls are without doubt quite thin enough to perform an exchange function, but is this sufficient evidence to state that this exchange of gases actually happens? Following the work of J.M. Soum in the 1890s it seems quite obvious that gaseous exchange, at least to any significant degree, does not occur. The air sacs have a very poor blood supply and the walls are smooth and do not possess the folded surfaces necessary to give the large surface area over which gaseous exchange may occur. Soum performed what for his period was a very sophisticated experiment, during which he introduced carbon monoxide into the air sacs of birds. He also blocked the connections between the air sacs and the rest of the ventilation system. If the air sacs were organs of gaseous exchange then the birds would obviously have died as a result of carbon monoxide poisoning. This did not in fact happen and therefore air sacs could not be the site of gaseous exchange. The fact remains, however, that air sacs do show considerable changes in volume during the ventilation cycle and so the 'bellows' theory seems valid. Let us now go on to see how the air actually flows in the avian lung.

## The Flow of Air in the Lung

Knut Schmidt-Neilson working at Duke University in the early 1970s used two techniques to investigate the path of air through the avian lung. Firstly he used a foreign gas as a marker (labelled with a radioactive isotope) and examined the time of its arrival at various points within the system. A second, but it seems to me a rather more insensitive, method was to surgically insert tiny electronic probes responsive to oxygen flow, at various points along the elaborate passages. Using an ostrich, because of its slow rate of ventilation, the Duke team discovered a rapid increase of oxygen in the posterior sacs near the conclusion of inhalation, showing that the initial flow was directly into the posterior sacs. In contrast, the oxygen did not appear in the anterior sacs until a full cycle later, i.e. at the end of the second inhalation, and that furthermore this oxygen came from the lungs. The mechanism of ventilation is now becoming clear. The entrance of air is due to an increase in volume of the body cavity brought about by movement of the sternum and ribs; the resulting reduction in pressure causes air to rush in via the nostrils and trachea. It then moves straight through the lung, bypassing the anterior sacs and going directly into the abdominal sacs; some air is pushed into the posterior thoracic sacs and also into the parabronchi. During expiration the body cavity is reduced in volume and air is forced from the posterior sacs into the lung. During the next inhalation this air continues to move along the parabronchi and into the anterior sacs, making room for more inspired air. Thus the anterior sacs serve as reservoirs to hold air rich in carbon dioxide prior to exhalation.

Thus the bird has the great advantage of continuous through-ventilation of the gas exchange areas and there is little 'dead space' which could dilute the incoming oxygen. Oxygen diffuses from the parabronchi into the air capillaries which are surrounded by a profuse network of blood capillaries. The air capillaries of a bird are some 3-10 microns wide as compared to a range of 30-1,000 microns for mammalian alveoli. Thus gaseous exchange in birds is very rapid, as indeed it must be in such active creatures.

## The Heart and Blood System of a Bird

Achieving gaseous exchange as smoothly as possible is one thing—the province of the heart and circulatory system is quite another and there is a huge strain imposed upon this system to move oxygen and other much needed materials rapidly into areas of high metabolic activity such as working muscles, and to remove carbon dioxide and other waste substances. Therefore the bird heart has evolved into a large and powerful organ with rapid muscular contractions. In both birds and mammals it is generally true to say that the smaller the species is, then the larger will be its relative heart size, and the quicker will be its heart rate. It is also generally true that gram for

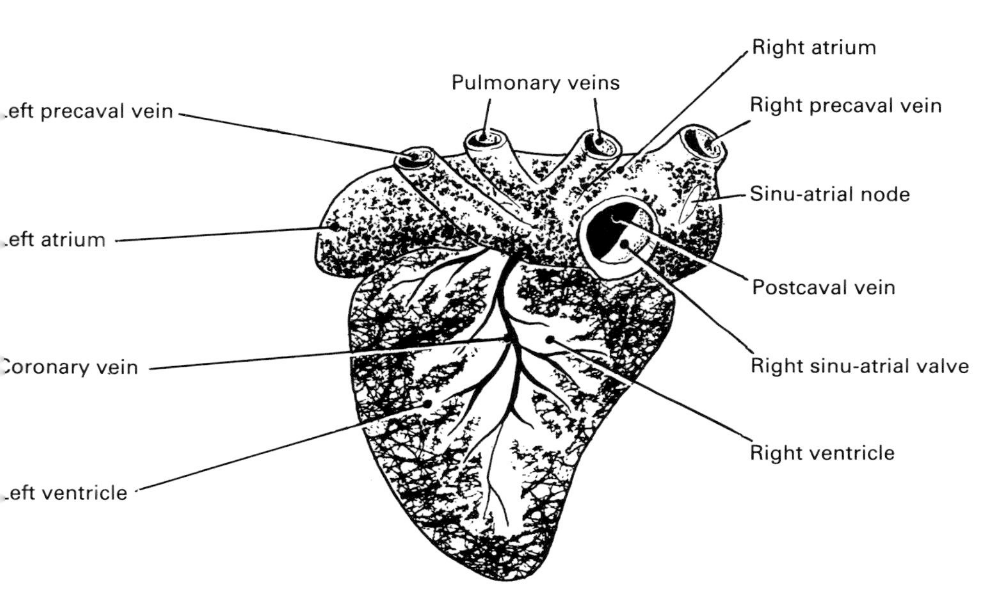


**Fig 73** The avian heart showing entry of the great veins

gram of body weight birds have larger and faster hearts than mammals. In man, for example, the heart weight is 0.42 per cent of body weight whilst the pulse at rest averages 72 beats per minute. In the house sparrow the heart is 1.68 per cent of body weight and the pulse at rest is 460. In the very tiny ruby-throated hummingbird these figures rise to 2.37 per cent and 615. The respiratory rates also show a similar trend, varying to some extent with the age of the bird; generally speaking, however, the rate is inversely proportional to body size, so that the smaller the bird, the faster its respiratory rate. For a domestic fowl the rate is 20 per minute, for the pigeon 26 and for the starling the rate rises to 84.

THE BIRD HEART

This consists of four chambers, two receiving vessels called atria, and two synchronised pumping areas called ventricles. This is similar to that of a mammal but there are many differences in the two systems and some of these will become apparent as we proceed.

Blood deficient in oxygen enters the right atrium through three huge veins called venae cavae. The blood passes into the right ventricle and is then pumped out via the pulmonary arch. From this arise two pulmonary arteries which carry blood to the lungs where gaseous exchange takes place. It is returned to the heart not in two blood vessels as in mammals, but by means of four huge pulmonary veins. This oxygenated blood enters the left atrium, then passes into the left ventricle prior to being pumped out under high pressure via the right systemic arch into the arterial system feeding the body.

ARTERIAL SYSTEM

The systemic arch bends over to the right and gives off paired innominate arteries, each forking into a carotid artery feeding the head, and a huge subclavian artery (*see* Fig. 74). This divides into two, producing a brachial artery to the wing, and an understandably large pectoral artery supplying the vital flight muscles. The systemic arch then curves around the right bronchus and behind the oesophagus, and now becomes known as the dorsal aorta. Just below the heart the coeliac artery leaves the aorta and then divides into two branches called the gastric and hepatic arteries. The small intestine is supplied by the anterior mesenteric artery. In birds there are three pairs of renal arteries (there is only one pair in mammals). The first pair arise directly from the aorta, but the second and third pairs, and the femoral arteries, lead off, these last carrying blood to the legs. The femorals arise further forwards than is the case in mammals—obviously related to the anterior position of the hind limbs. Behind the kidneys the paired iliac arteries and the posterior mesenteric artery arise at the same level. A small caudal artery leads towards the tail. Thus the blood has been carried to the far extremes of the body. It is now the responsibility of the venous system to return the blood, which now contains high concentrations of carbon dioxide, to the heart and thence to the lungs.

VENOUS SYSTEM

Each of the two superior venae cavae receives a jugular, brachial and pectoral vein and are thus huge blood vessels (*see* Fig. 75). In the anterior part of the neck, the jugulars are joined by a structure called the anastomosis, a precaution to ensure that if the flexible neck is twisted enough to constrict the jugular vein on one side, the blood can return to the heart on the other side. This is particularly important in long-necked birds such as swans or herons.

Blood from the tail returns in the caudal vein which divides into three branches. The two outermost branches run in front of the kidneys receiving the iliac, sciatic and femoral veins and receiving a few small branches from the kidneys. The two branches unite above the kidneys to form the inferior vena cava, which leads forward to the right atrium receiving the epigastric and hepatic veins from the liver. The middle branch from the caudal vein proceeds forwards as the coccygeo mesenteric vein and joins the hepatic portal. Blood from the gut goes to the liver in the hepatic portal vein, as in all vertebrates. In birds it is noteworthy that the renal portal system which is so important in mammals is practically non-existent, and most of the blood from the posterior region of the body is returned to the heart as quickly as possible. This is almost certainly because the large pectoral muscles demand a huge blood supply and a speedy return of blood from less essential regions has proved an evolutionary advantage.

One final word on the blood of birds concerns the red blood cells

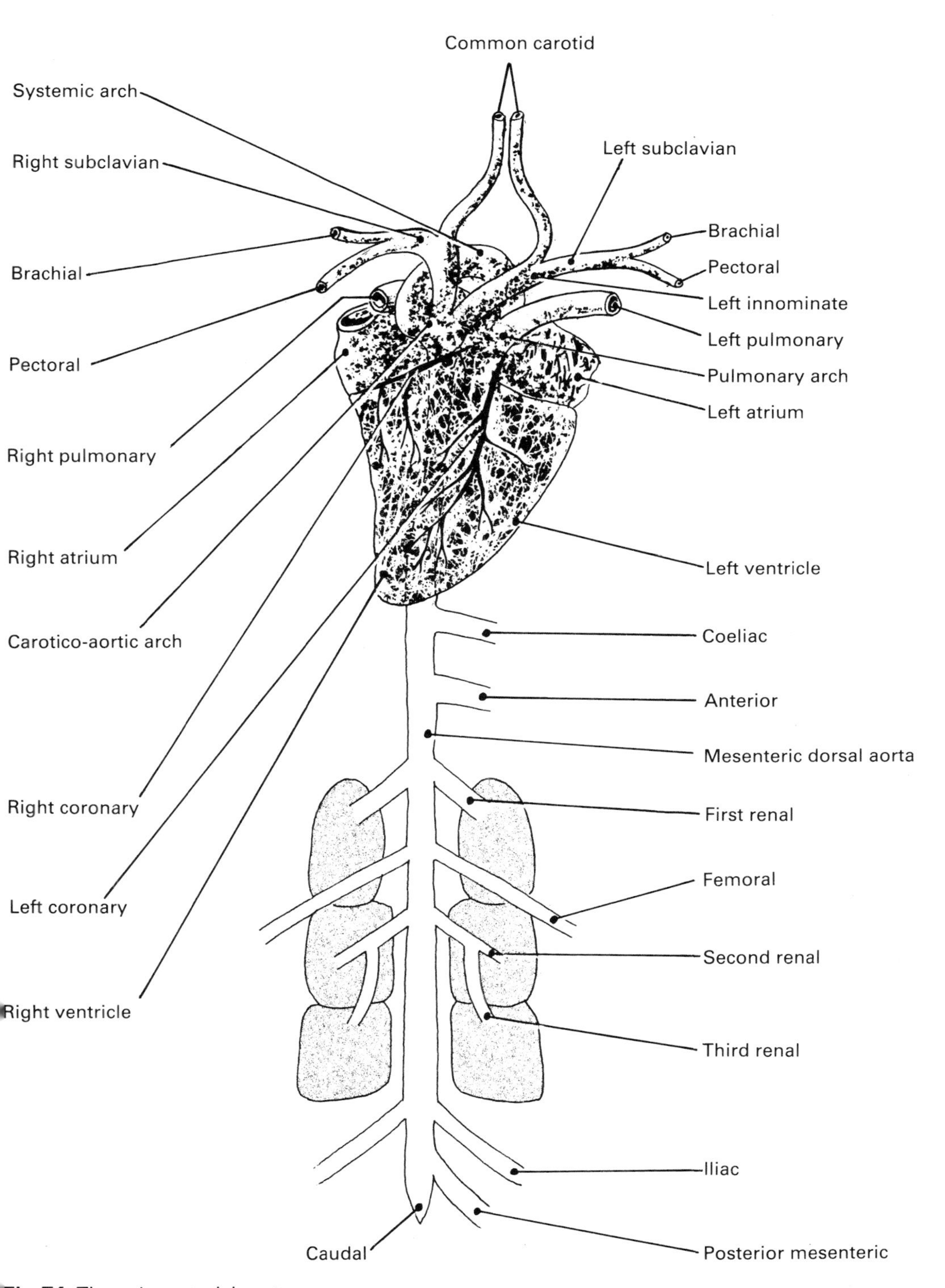


**Fig 74** The avian arterial system

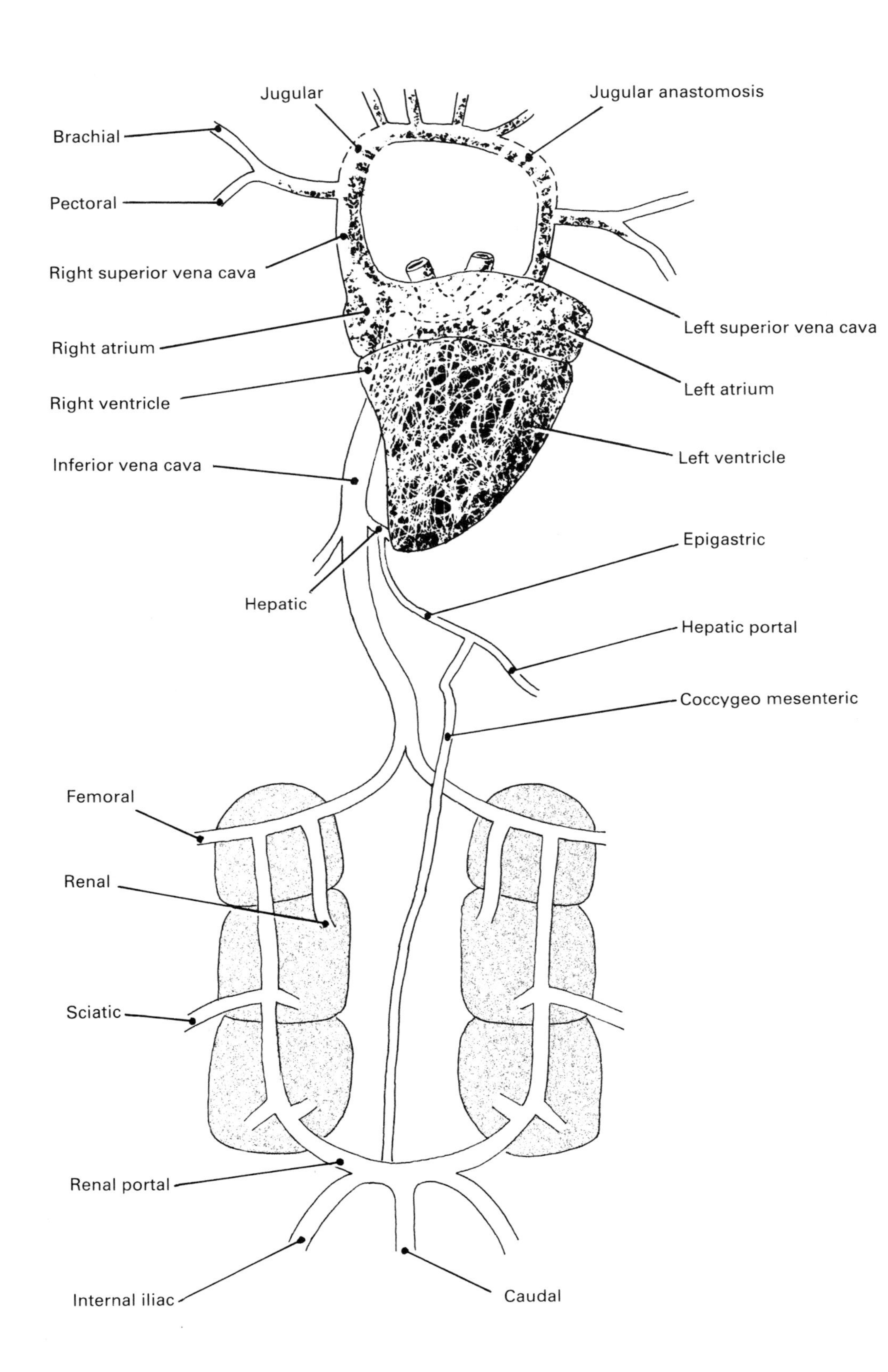
Jugular
Jugular anastomosis
Brachial
Pectoral
Right superior vena cava
Left superior vena cava
Right atrium
Left atrium
Right ventricle
Left ventricle
Inferior vena cava
Epigastric
Hepatic
Hepatic portal
Coccygeo mesenteric
Femoral
Renal
Sciatic
Renal portal
Internal iliac
Caudal

(erythrocytes) themselves. These contain the red pigment haemoglobin which is involved in oxygen transport. In reptiles such as the lizard the blood cells are ovoid and nucleated; in mammals the cells are disc-shaped and have no nuclei when mature. The bird resembles its reptilian ancestors in retaining ovoid nucleated erythrocytes, but in the development of a sophisticated blood system and a highly efficient method of ventilation birds have shown great advances in their design for living.

**Fig 76** The venous system copes efficiently even for long-necked birds such as this whooper swan *Cygnus cygnus*

**Fig 75** (opp.) The avian venous system

*Chapter Six*

# Feeding

Like all other avian systems the alimentary canal has become adapted to the demands imposed by flight. The head, compared to that of a mammal, is quite small, and the luxury of huge heavy muscles to chew food simply cannot be afforded. Teeth composed of the dense materials dentine and cement were dispensed with early in the evolution of birds to be replaced by a much lighter, horny bill.

In order to retain the essential high rate of metabolism birds need to satisfy two requirements. Firstly the actual digestive process must be swift, and secondly the main organs concerned with dealing with the food must be positioned close to the bird's centre of gravity so that the weight interferes with balance as little as possible during flight.

The best way of understanding the mechanisms involved is to follow the path taken by the food from its ingestion at one end to the elimination as waste material at the other (*see* Fig. 80).

## The Bill

In the primitive bird, the bill was composed of quite a number of separate plates which during the course of evolution became fused together, but some traces of the original plates can still be detected in the ostrich and some of the skuas. In the Procellariformes such as the shearwaters and petrels the plates still remain separate, as a look at the bill of the fulmar *Fulmarus glacialis* clearly shows (*see* Fig. 77).

**Fig 77** The fulmar's *Fulmarus glacialis* 'tubenose' may function as a means of measuring wind speed during flight or possibly as a primitive organ of smell

The bill is subject to great stresses, and the wear and tear upon it must be very considerable. Unlike the plumage, the bill, with one interesting exception, does not undergo a true moult; instead it grows throughout the life of the bird to compensate for the degree of erosion. Should damage occur to the bill it may grow abnormally and prevent the bird from feeding efficiently. The starling *Sturnus vulgaris* seems to be particularly susceptible to this phenomenon, and I once saw an individual with an upper mandible considerably longer than normal and shaped just like a corkscrew. The one European exception to the 'no-moult' rule is found in the puffin *Fratercula arctica*. During the breeding season the puffin develops extra brightly coloured plates on the bill. These are shed in the autumn when the bill becomes somewhat less bulky and very much less attractive (*see* Plate I, p. 137).

The bill of a bird is the principal organ used to search for food and so it is not surprising that the shape has undergone many variations during the course of evolution. As we saw in Chapter 1 the Galapagos finches have developed different bill shapes in order to exploit a specialised niche. Those species with the best adaptations will be the ones which get the major share of the available food, and will be the ones which survive the environmental pressures at the time. A look through any field guide will reveal great variations in bill structure from the tearing weapons of the birds of prey, the serrated bills of the fish-eating mergansers, the delicate structures of the insect-eating warblers, the wide gape of the nightjars to the tough seed cracking bills of the finches. The degree to which this specialisation goes can be seen by an examination of the more subtle variations in bill shape shown by groups of closely related birds enabling them to avoid direct competition. This is clearly demonstrated by the waders (*see* Fig. 78).

It is wrong to assume that a bird's bill is dead and insensitive; indeed it is well provided with a number of highly sensitive cells. The bills of the woodpeckers which probe into both natural and self-constructed crevices, and those of ducks which dive for sustenance into water which is often too murky for accurate vision are particularly well endowed with nerve endings. Waders such as the woodcock *Scolopax rusticola*, snipe *Gallinago gallinago* and curlew *Numenius arquata* probe into soft earth or sand and have bills covered with a sensitive tissue containing many sensory pits.

It is not only with regard to bill shape that we find variation; the avian tongue has also been subject to the pressures of evolution and has come up with some interesting adaptations. The tongues of the hummingbirds which feed mainly on nectar are especially long and sensitive. As one would expect, species feeding mainly on fish, such as the cormorants, have rough spiky tongues which resist the escape of struggling and slippery prey, whilst the searching probing tongues of woodpeckers are particularly sensitive. In the green woodpecker *Picus viridis* a further adaptation is the considerable enlargement of the salivary glands situated on the floor of the buccal cavity. The secretions produced by these structures keep the tongue sticky enough to enable insects to be easily captured. The great spotted woodpecker *Dendrocopos major* even has a barbed tip to the tongue which further increases its efficiency, allowing it to 'harpoon' its prey. The jay *Garrulus glandarius* also includes a high proportion of insects (mainly ants) in its very varied diet, and has well-developed salivary glands and tongue as do the insect-eating hirundines.

In some waders, particularly snipe, the nasal bone is not attached to the base of the skull and this enables the tips of the bill to open to grab prey without the whole bill filling with mud. In the puffin *Fratercula arctica* an extra bone has been retained thus allowing the mandibles to 'open parallel' and a row of fish can be held without dropping any.

Birds do not possess the soft palate which assists us mammals to swallow easily, but instead they must throw their heads backwards to persuade the food to enter the oesophagus; during the period of swallowing the entrances to the ventilation system (*see* Chapter 5) are closed by reflex actions.

**Fig 78** This comparison of waders feeding shows how their differing bill lengths allow them to extract food items at different levels. The worms Nereis and Arenicola, and the bivalve Scrobicularia can be reached only by the bird waiting to catch them close to the surface

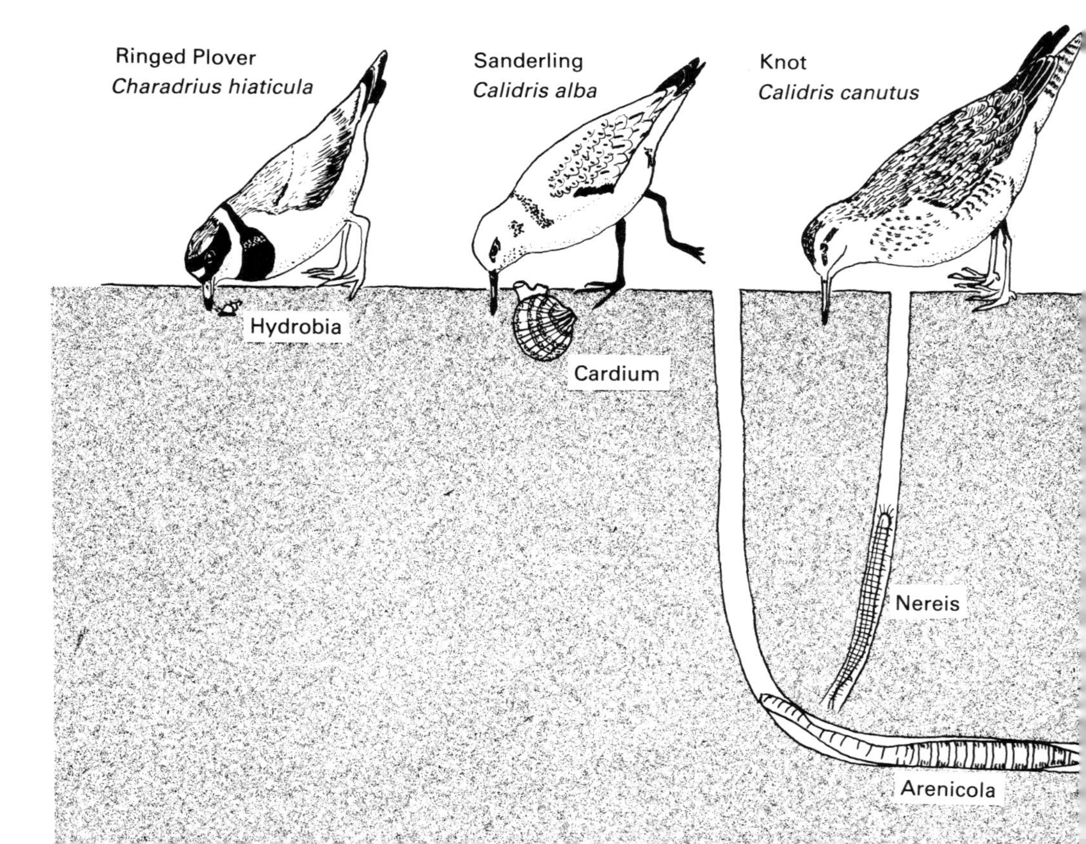

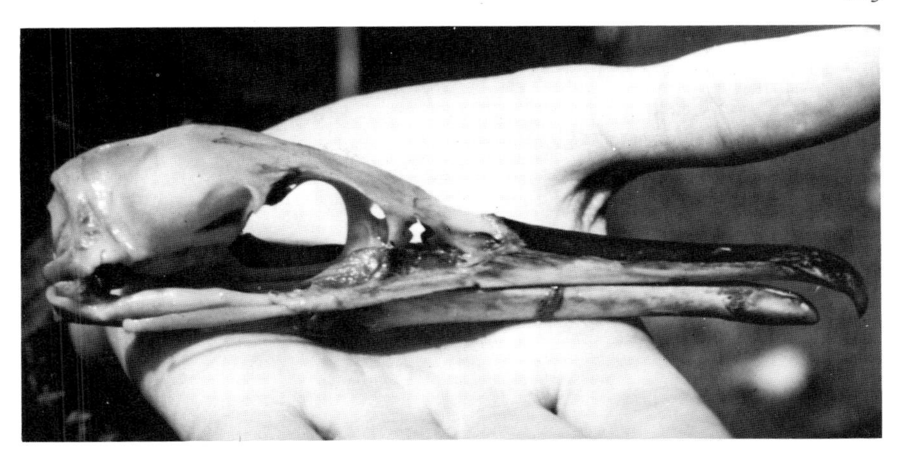

**Fig 79** The skull of a cormorant clearly showing the hooked bill

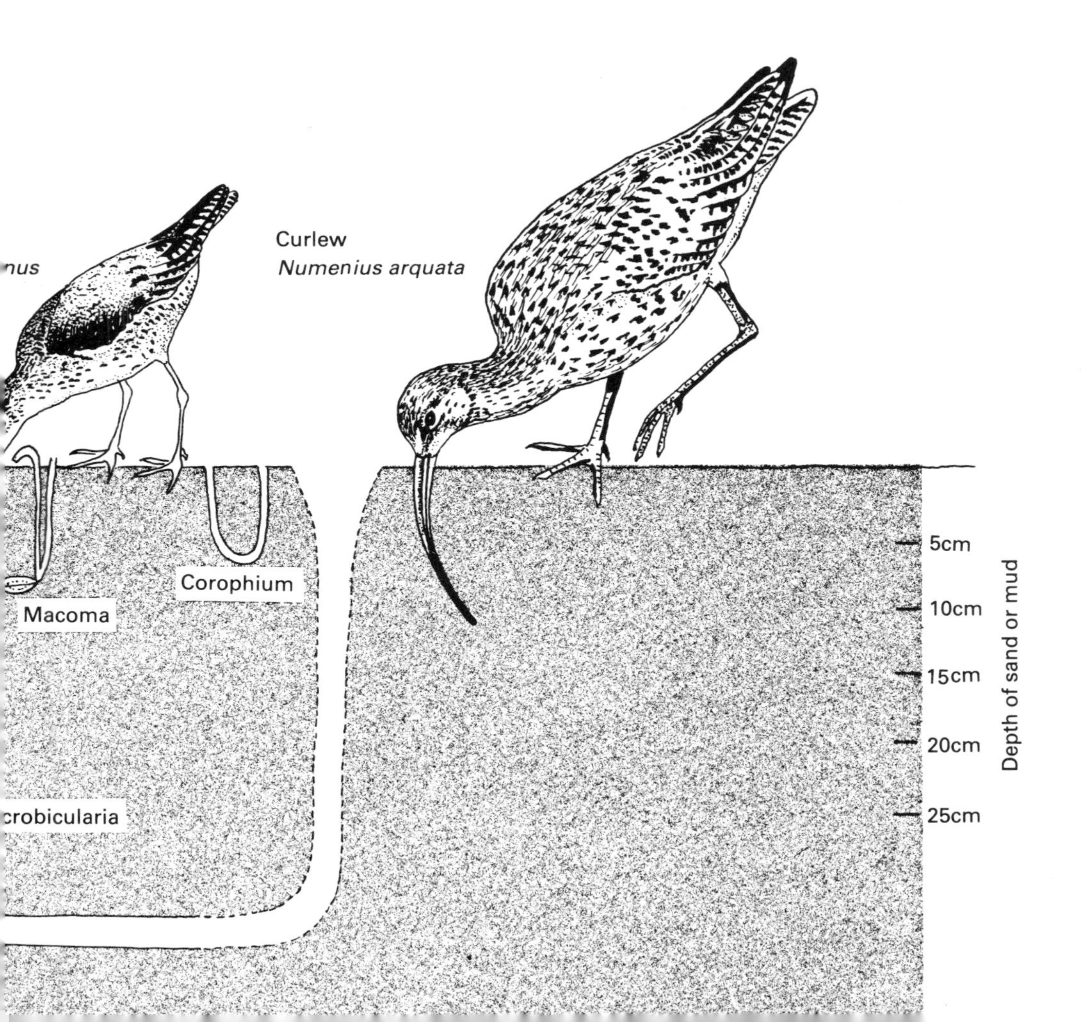

## The Oesophagus

In mammals the food is torn apart and chewed by means of the teeth; it is then rolled into a rounded bolus prior to being swallowed. The diameter of the oesophagus can therefore be quite narrow and smooth. In the case of birds their prey is often covered with fur, feathers or sharp spines, and in any case is swallowed whole, and so the oesophagus needs to be correspondingly wider and more muscular than that of a mammal. This is especially true of birds of prey and fishers like the cormorants and the kingfishers, and in some of these part of the oesophagus has become dilated to serve as a storage organ. In some birds this process has gone one step further and resulted in the evolution of a structure called the crop.

## The Crop

This organ is particularly well-developed in game birds which consume comparatively large volumes of vegetable material often obtained from areas abounding with hungry predators. Like their herbivorous mammalian equivalents such birds have evolved a storage vessel, the contents of which may be digested later in the safety of a convenient roost. With the notable exception of the South American hoatzin *Ophistocomus hoazin* it is not used for digestive purposes. It is in the pigeons that the crop reaches its peak of perfection; not only does it provide ample storage capacity, but during the breeding season the lining becomes soft and produces a milky substance on which the young squabs are fed during a period from the eighth to the sixteenth day following hatching. Doubts have been expressed as to whether this is a true milk in the accepted mammalian sense, but the constitution of the two is remarkably similar except that the sugar content of pigeon milk is well below that of the milk of mammals. The production is controlled by the hormone prolactin directed from the pituitary gland—precisely the same mechanism as that triggering the production of mammalian milk. Apart from the pigeons, the avian world has ignored the virtues of milk.

The oesophagus leads into the stomach and it is in this region that the bird's digestive energies are largely channelled.

## The Stomach

Any efficient digestive system, including our own, must include a mechanical factor to break down large pieces of food material, and a chemical factor which works at a molecular level so that carbohydrates which provide energy, proteins which give body building materials, and fats which act in a storage capacity can be absorbed into the blood system to be transported around the body. The bird's stomach is divided into two regions: the proventriculus; and the gizzard which is purely a mechanical device.

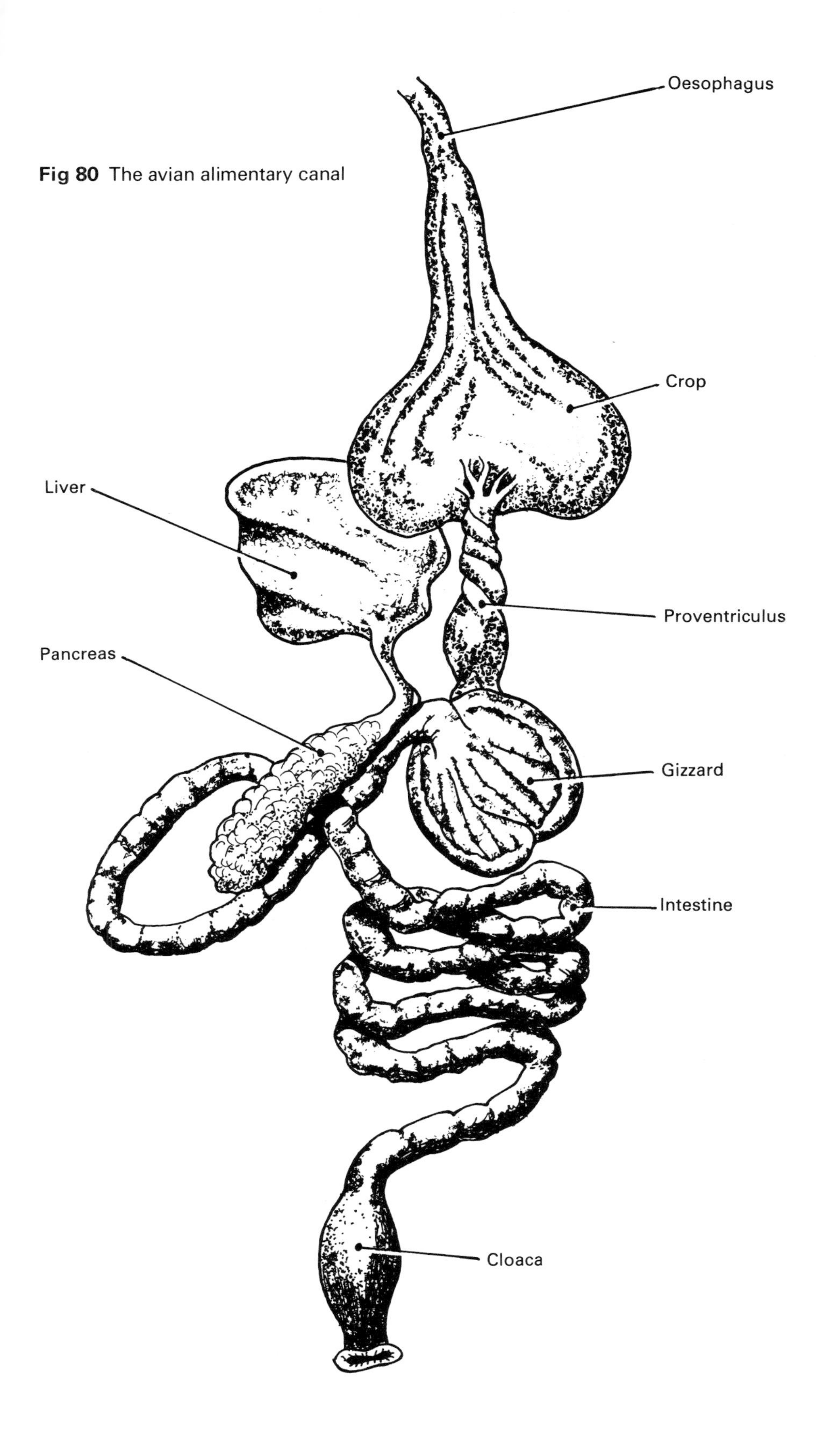


**Fig 80** The avian alimentary canal

THE PROVENTRICULUS

This is extremely glandular though the actual degree of development depends a great deal on the diet; in the Procellariformes, however, especially the fulmar *Fulmarus glacialis*, it has a unique function—that of protection. Each bird produces about 200 cc (7 fl oz) of stinking, viscous, reddish-coloured oil which is directed with unerring accuracy and considerable force at any predator. In the days prior to the mineral oil industry the native population of St. Kilda in the Outer Hebrides had a thriving trade in fulmar oil which was used both for cooking and as fuel for lamps.

In other predatory birds with gargantuan appetites such as gulls, skuas and herons the proventriculus provides additional storage space, but its main function remains that of chemical digestion, and in all birds it is here that the food becomes thoroughly mixed with digestive juices, especially the protein enzyme pepsin. Once this has been accomplished the partially digested food is passed into the gizzard.

THE GIZZARD

It is in this region that great variation occurs from one family of birds to another. It is in those whose diet consists of vegetable matter that the gizzard is the most highly developed. Strong circular muscles are present in the walls which are greatly toughened by a lining made of a substance called kaoilin, and between these the food is ground in the manner of old time millstones. As if this were not sufficient in itself many birds deliberately swallow sharp stones and grit in order to assist this grinding action. This system, then, is the bird's answer to the mammal's teeth.

In contrast to this grinding mill the carnivorous birds such as cormorants, saw-billed ducks, owls and diurnal raptors have very poorly developed gizzards simply allowing the digestion begun in the proventriculus to continue for an extra period of time. The gizzard does have a very vital part to play in the life of these predatory birds however, since should any indigestible bones proceed beyond this point into the delicate soft-walled intestines designed for absorption of digested food then permanent and even fatal damage could ensue. Thus the gizzard has various strategems which vary from family to family to block the passage of fur, feathers and bones. These materials are rolled into a pellet and forcibly expelled through the mouth. Any bird is capable of getting rid of unwanted material in this way, but the pellets are obviously much larger in birds of prey and are used extensively by some ecologists who are able to identify the precise nature of the diet of the species under review without having any reason to kill the bird in order to analyse the gut content.

Once the indigestible material has been expelled and time allowed for efficient digestion to have occurred the sphincter muscle at the far end of the gizzard relaxes and the material is squeezed by a peristaltic wave of contraction into the next part of the alimentary tract, the intestine.

## Intestine

As in mammals, the bird's intestine has two basic functions to perform, namely to complete the digestion begun earlier in the digestive tract and then to provide a large surface area through which the final products of digestion may be absorbed into the blood stream. In contrast to mammals, birds cannot allow the soft parts of the digestive tract more or less unrestricted power of movement; the restrictions imposed by flight have resulted in the intestines being held more firmly in place.

### DIGESTIVE SECTION

Regarding digestive fluids the two classes Aves and Mammalia are similar, the first part of the intestine of both receiving digestive fluids from the pancreas and from the liver.

The pancreas is very important in birds, producing carbohydrate-, protein- and fat-digesting enzymes. Unlike mammals, birds have no extensive digestion in the mouth which results in carbohydrates being partially digested prior to swallowing, and to compensate for this the pancreas is more substantial. This is also true of the liver which consists of two lobes, each having its own duct leading to the intestine rather than a single connecting duct as in mammals. A further point of difference is that the bile is acid in birds and alkaline in mammals.

### ABSORPTION SECTION

Although it is held more firmly in position, the function of this region is similar to that of mammals, and the end products of digestion are absorbed by a very rich network of blood vessels. These end products are glucose (from carbohydrates) amino acids (from proteins) and fatty acids and glycerol (from fats). Little variation in intestinal structure is seen until we reach the last section close to the junction with the rectum. Some species develop a pair of caeca at this point. In the grouse family they are particularly well-developed and play an important part in the absorption of protein and also, to a certain extent, water. The most important function of these caeca, however, is to serve as the only site for the digestion of cellulose, a vital thing for any herbivorous organism. No bird (or mammal) can actually digest cellulose by means of its own chemical enzymes, but present in these intestinal caeca are vast numbers of bacteria which can effect this breakdown. It seems that the bacteria are unable to survive for long outside the gut of the bird and they allow a share of the breakdown products of cellulose in exchange for shelter, a relationship known as symbiosis.

Such development reaches its peak in birds such as grouse and ostriches, whilst in birds such as swifts and hummingbirds they are very much reduced and in raptors and fish-eaters the caeca are absent altogether. By whatever method, a particular species will, by the time the end of the intestine is reached, have extracted the maximum food material from its chosen diet. What remains passes onwards into the rectum.

## The Rectum

The function of the rectum, which is quite short, is to serve as a site for the absorption of water into the blood-stream and to conduct the solidified material along to the cloaca which is the joint opening shared by both the alimentary canal and the sexual ducts, again a point of contrast with the mammals where the two openings are separate.

This, then, is how the bird handles the food which it manages to find; the next point to discuss is the speed at which it can be utilised. Obviously the more active the bird is, the more food it will require. The demand will rise during flight, cold weather, moult, migration and the breeding season, but as a general rule the smaller the bird, the higher its metabolic rate, and species such as wrens and warblers may need to consume over 30 per cent of their body weight of food each day. This figure drops to something in the order of 10–15 per cent in pigeons, whilst in the relatively sedentary domestic hen the figure may be under 3 per cent. This automatically means that the digestive system of a small bird needs to function more quickly than the larger species, but occasionally nature has evolved a short cut. In the flowerpeckers, Dicaeidae, for example, the diet consists of fruit rich in simple sugars which need little, if any, digestion. Here we find a bypass leading from the oesophagus directly into the intestine without passing through either the proventriculus or the gizzard, obviously a great saving in terms of time. Small birds are known to be capable of digesting berries in under an hour; this was shown by the simple method of feeding captive birds with a favourite item in the diet, having previously impregnated it with a coloured dye.

Anyone who has ever tried to look after a young bird will know how often they need to be fed, and this has sometimes led to the suggestion that birds must feed every few hours if they are to survive. This is stretching the truth a bit, but they certainly do not cope as well as mammals in times of food shortage; their metabolic rate has to be so much higher, however, that we can hardly expect anything else. Laboratory experiments have shown that the house sparrow *Passer domesticus* is capable of surviving without food for up to three days whilst the larger mallard *Anas platyrhynchos* may survive for 22 days. The avian fasting record without doubt belongs to the male Emperor penguin *Aptenodytes forsteri* which patiently incubates an egg throughout the worst two months of the long Antarctic winter, living only on the huge fat deposits stored beneath the skin.

Birds have two main stratagems to help them overcome periods of food shortage, the development of the crop assisting in the short term, and the laying down of large fat deposits for the long term (*see* Chapter 9). A supply of food is not the only thing required to keep the 'bird machine' working at maximum efficiency—a regular supply of water and a regular internal balance is essential. All birds take in water through the mouth, the usual method being to scoop up water in the bill, then to tilt the head allowing water to be carried into the gut by gravity. Pigeons are the only birds which can drink like a mammal, actually sucking water against gravity. Seabirds have a special problem in obtaining fresh water since they spend much of their lives hundreds of miles, in some cases thousands, from the nearest supply. They drink seawater and have evolved a pair of specialised salt glands above the bill which can excrete excess salt via the nostrils.

The very fast metabolism of birds produces high concentrations of waste products, and as birds have no sweat glands they depend upon their paired kidneys for water balance and these are much larger than those found in mammals. All vertebrate kidneys are based on a similar pattern, but those of birds resemble a reptile rather than a mammal kidney, there being very many more tubules which filter off the urine from the blood supply and this is also much more extensive than that of any mammal. Furthermore, birds do not excrete liquid, soluble urea, but solid uric acid, again showing their close relationship with reptiles. Urine does, however, reach the cloaca as a clear fluid, and here it is mixed with the faeces in a region known as the proximal sac. It is here, too, that the water is reabsorbed into the blood stream, enabling the bird to make maximum use of every drop and thus allowing long journeys across hostile environments to be attempted.

**Fig 81** Mallards *Anas platyrhynchos* are vegetarians: a weed-covered pond is an ideal habitat

*Chapter Seven*

# The Nervous System

The nervous system of any vertebrate can be considered in three parts: the central nervous system, the sense organs, and the hormonal or endocrine system.

## Central Nervous System

This consists of a brain and spinal cord connected to all parts of the body by nerves, which are made up of large numbers of nerve cells called neurones, arranged rather like electrical wires within a cable, and which actually conduct nervous impulses. They are of two basic types: the sensory neurones conduct messages received by the sense organs to the brain; motor neurones transmit the message from brain to muscles. Thus a blue tit *Parus caeruleus* may see or hear a sparrowhawk *Accipiter nisus*. The image of the predator speeds along the sensory neurones to the brain which registers the urgency of the situation. Messages flash back along the motor neurones to the flight muscles and to the legs and the blue tit hops into the air and flies away from the danger.

## Sense Organs

These are concerned with vision, hearing, smell, taste and touch. We must remember that we make this five senses classification on the basis of our own system and should appreciate that birds may well have some other system of monitoring their environment which we do not need—a sort of sixth sense.

### SIGHT

The survival of a bird depends much more upon efficient vision than anything else, and a high percentage of the 'evolutionary budget' so far as the nervous system is concerned has been poured into improving this one sense. A casual glance at a bird's eye does not give any idea of the actual size of the eyeball; it is only on dissection that this can really be appreciated and it is surprising how little else is left after the eyes have been removed (*see* Fig. 82). A bird could almost be described as a pair of flying eyes and even the muscles which move them have been sacrificed. It is therefore much more difficult for a bird to move its eyes than it is for a mammal which has larger muscles and smaller eyeballs. Gulls and also the herons can move their eyeballs at least to some degree, but most birds have solved the problem of seeing through a wide arc by evolving long or extremely flexible necks, a feature which is most obvious in the owls (Strigiformes); these birds of the

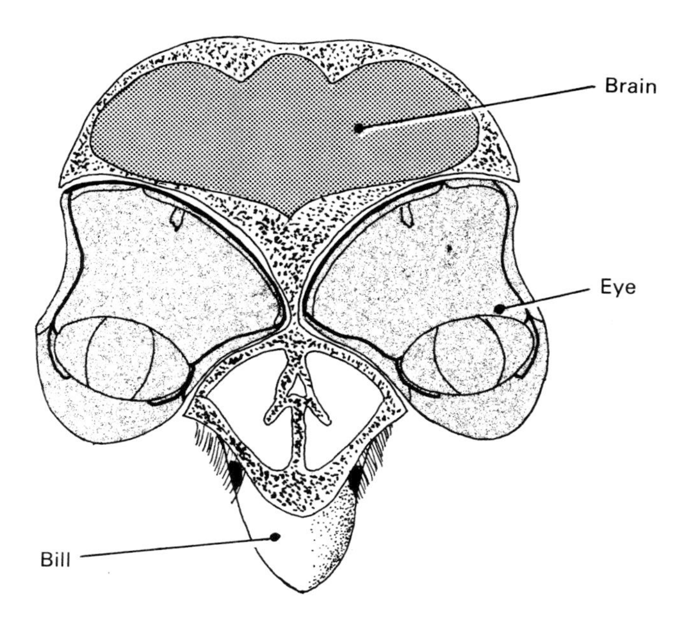


**Fig 82** This skull of an eagle owl *Bubo bubo* shows the large size of the avian eye compared to the total volume of the cranium

night can turn their necks through an angle of at least 180° and it has been humorously suggested that if you walk slowly around a roosting owl it will follow your progress and eventually wring its own neck!

The bird's eye is of the standard vertebrate pattern, but with a number of important variations. A lens focuses the light energy onto a sensitive retina which lines the eyeball; this retina is composed of two different types of cell so named because of their shape—the rods and cones. The rods are much more sensitive to low intensity light but do not produce a clear image, whilst the cones provide not only a sharp image, but also give colour vision. The eyes of owls have many more rods than cones—in other words they have sacrificed some degree of colour vision in order to be able to hunt in semi-darkness. A look at cross sections drawn of the eyes of a bird and mammal (*see* Fig. 83) will reveal some significant differences. In the bird's eye we find a structure called the pecten intruding from the back of the eyeball into the vitreous humour and which is a folded membranous structure very well supplied with blood vessels. The pecten shows many variations in shape, being particularly well-developed in diurnal birds. There has been a great deal of speculation regarding the precise function of the pecten, including the idea that it is an erectile organ which by changes in its shape can alter the pressure within the eyeball, thus affecting the shape of the lens and allowing the eye to accommodate any changes occurring as animals move closer to the bird or it approaches nearer to other objects. Another suggestion is that the

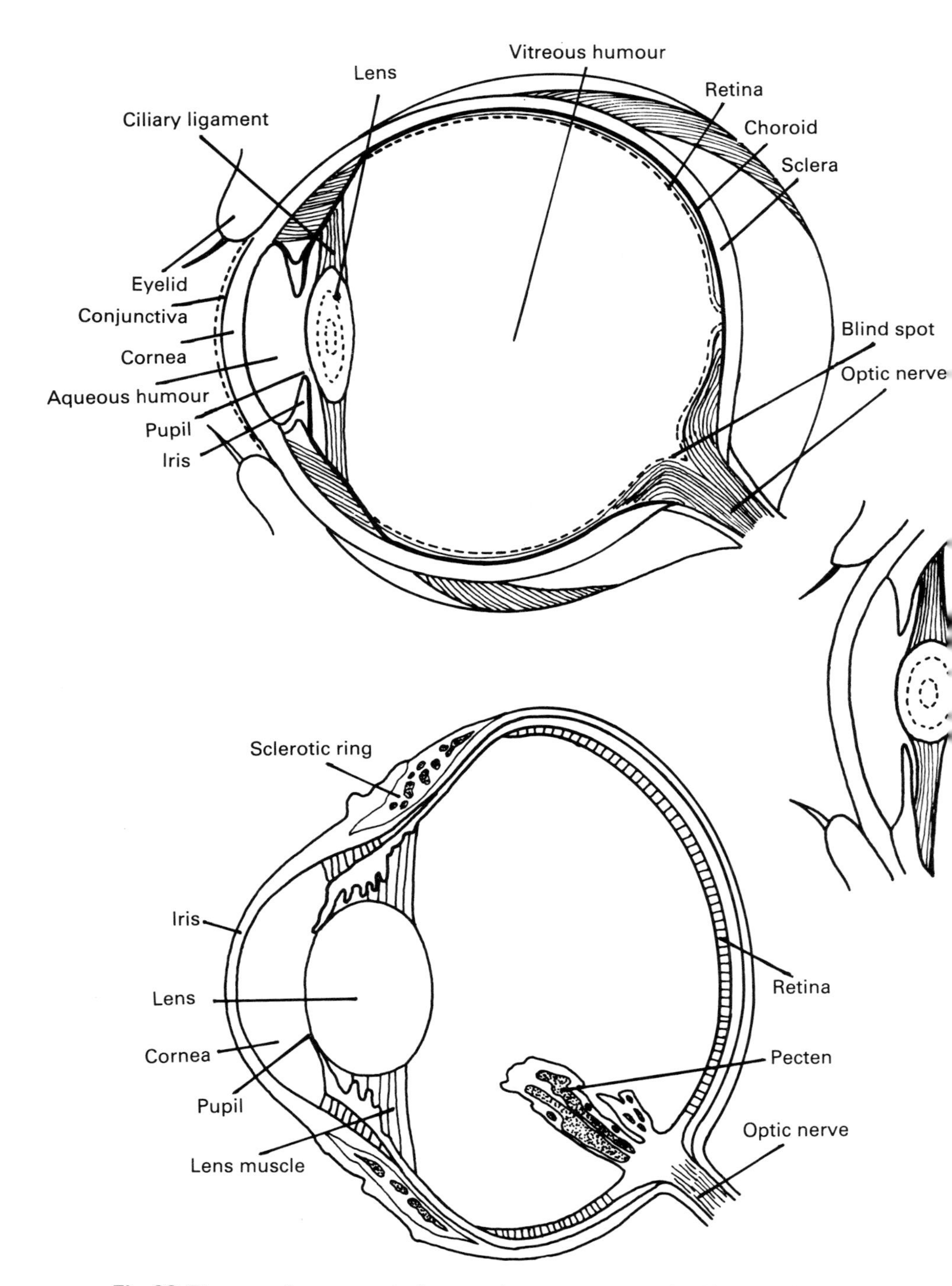


**Fig 83** The eye of a mammal *above* and *centre* compared with that of a bird. Note that whereas the ciliary ligament in the mammal eye is used to alter the shape of the lens, in the bird's eye the alteration is far smaller and depends on the pressure in the eye

shadow of the pecten cast upon the back of the retina might well provide the impression of any movement in the environment—a very useful thing for both hunter and hunted. The main weight of opinion, however, seems to suggest that the pecten is a sort of supplementary feeding organ, providing the retina with materials which diffuse though the vitreous humour. From an evolutionary point of view this makes sense since a similar structure called the arteria centralis retinae is found in reptiles. A final suggestion regarding the function of the pecten suggests that it may be concerned with navigation but a great deal of very difficult research is yet required in this particular field.

A further contrast between the avian and mammalian eye concerns the fovea: in mammals this is a small indentation in the centre of the retina and which contains only cones. Around this there is a very high concentration of sensory cells and this means that at the fovea very sharp images can be formed with an accurate colour balance and transmitted quickly via the optic nerve to the brain. In birds there may be one or more foveae, but they do not have precisely the same function as in mammals. They may well work in conjunction with the pecten to enable a bird to detect movement against an otherwise featureless background, such as the sky or over the sea, an ideal faculty for a flying animal to have.

Let us now come to the question of just how much of the environment can be detected and discuss the quality of the image obtained. Again, as we would expect, differences between birds and mammals occur, the eyeball being flatter in birds which means that it is able to maintain the whole visual field in focus at the same time. A mammal doesn't have this advantage and must concentrate its attention on one small area at a time; from these it constructs a composite picture. Furthermore a bird's eyes tend to be situated on the sides of the head, and this means that it can cover a much wider field.

Thus far it would appear that all the visual advantages are with the birds. Let us now redress the balance. With a few exceptions, notably in the owls, the eyes are so wide apart that there is not very much of an overlap in the images received from the two eyes and this causes a lack of perspective, a very serious deficiency. Put simply this means that birds have mainly two dimensional vision compared with the three dimensional sight of mammals. Next time you get a chance to watch a common passerine bird feeding, notice how it continually cocks its head on one side. Many old-time naturalists wrote about birds listening for earthworms sliding their way upwards through the soil, but what they are actually doing is compensating for their lack of judgment of distance by looking at prey first from one angle and then the other. This is making use of the physical properties of parallax and it allows the bird to build up a three dimensional picture and thus enables it to judge distances more accurately.

This lack of a three dimensional image must result in problems for birds which hunt on the wing and several stratagems have been evolved in an attempt to solve these. In many birds of prey and also in some divers, marks run from eye to bill and seem to function like sights or guidelines, allowing the bird to aim at its prey. The head is also moved from side to side so that distance can be judged by parallax. Insectivorous birds like the hirundines have tackled the problem by evolving two foveae in each eye. The sharp images formed on the temporal and central foveae produce very efficient binocular vision. It seems that the swallow *Hirundo rustica* has the best of

**Fig 84** The nictitating membrane seen here in the woodpecker *Dendrocopus major* is often drawn across when a bird feeds its young as it affords some protection against the eagerly grabbing beak of the young.

both worlds because it is also able to use both eyes independently to secure monocular vision. Insectivorous birds which need to keep their heads fairly still whilst eating need to be able to detect any swift movement on their horizon which could mean the imminent presence of the dreaded peregrine falcon *Falco peregrinus*. In this case the fovea is in the form of a long narrow band, rather than in separated spots with less efficient areas in between.

The actual position of the eyes will also determine the degree of binocular vision, but in pigeons it is only about 20° out of a total field of 340° (this compares with the human figure of 140° overlap out of a field of 180°). Owls have their eyes situated more or less frontally, and their bills are deflected downwards which clears their field of vision; this produces a rather narrow visual field of about 110° but 70° of this is binocular.

No account of bird vision would be complete without mention of the nictitating membrance. A bird's eye has an upper and a lower eyelid, the lower one being much more developed and freely moveable than the upper. There is also a third eyelid, the nictitating membrane (membrana nictitans) which is normally hidden in the inner corner or medial edge of the eye, but can extend to pass over the eye like an automobile windscreen wiper. It is controlled by two special muscles and also has its own lubricating gland. The whole membrane is transparent which must be very useful as a windshield when the bird is in flight. In some divers, coots, ducks and auks the nictitating membrane is shut down over the cornea of the eye whilst the bird is diving and thus acts as a sort of contact lens. In man the nictitating membrane is present in the corner of the eye as the tear gland.

### HEARING

The 'ear' first evolved in fish not as an organ of hearing, but rather as one concerned with balance, a function which is retained to this day even though sound is such a vital thing in our lives that we often forget how and why the organ began. The ears of a bird occupy a similar position relative to the eyes as they do in mammals, but the pinna, a sort of biological ear-trumpet, is always lacking and the long eared owl *Asio otus* is incorrectly named since it has no prominent ears, its 'ears' being merely ornamental tufts concerned with display. The actual opening to the ear is covered by a special set of feathers called auriculars (or ear coverts). These feathers normally lack barbules and will therefore give protection to the ear without interfering too much with the sound.

As in a mammal the bird's ear can be divided into the three regions of outer, middle and inner ear (*see* Fig. 85). The outer ear without a pinna is a simple tube through which sound waves pass on their way to strike the ear drum and set it vibrating. The middle ear consists of an air-filled cavity which acts like the resonating box of a stringed instrument and across it stretches a long slender bone called the columella which transmits and amplifies the vibrations of the ear drum (tympanic membrane). In mammals the columella is replaced by three separate bones called the hammer, the

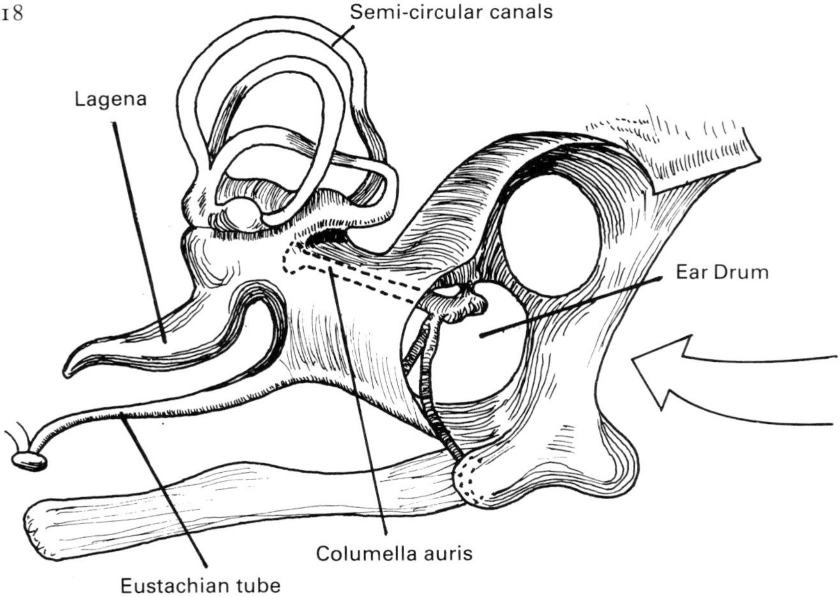

Semi-circular canals
Lagena
Ear Drum
Columella auris
Eustachian tube

**Fig 85** The internal structure of the avian ear, showing the single bone columella auris which in mammals is replaced by the three separate ear ossicles

anvil and the stirrup. The avian inner ear is occupied by a complicated arrangement of fluid canals called labyrinths, which are bathed also in fluid. Five regions can be recognised of which two, the utricle and the semicircular canal system, are concerned with balance and three, the sacculus, cochlea and lagena are concerned with hearing. The cochlea transforms sound waves into nerve impulses which are passed to the brain via the auditory nerve. The lagena contains tiny granules of calcium carbonate which are called otoliths or ear stones. It is thought that this area is responsive to low frequency sounds and the rest of the cochlea and the sacculus are concerned with the reception of high frequencies. In 1951 Beecher suggested that the otoliths might form a gravity system, capable of reporting changes in head position. There also seems to be a positive correlation

Table 7
SHOWING HEARING RANGES OF SELECTED SPECIES

*Units given in Hertz (cycles per sec)*

| Species | Lower limit | Maximum sensitivity | Upper limit |
|---|---|---|---|
| *Anas platyrhynchos*–mallard | 300 | 2,000–3,000 | 8,000 |
| *Phasianus colchicus*–pheasant | 250 | ? | 10,500 |
| *Asio otus*–long eared owl | 100 | 6,000 | 18,000 |
| *Sturnus vulgaris*–starling | 100 | 2,000 | 15,000 |
| *Fringilla coelebs*–chaffinch | 200 | 3,200 | 29,000 |

between the length of the cochlea and the complexity of the song pattern, but much work is needed before we fully confirm either of these two hypotheses.

It is interesting to consider what the range of hearing in birds actually is; some work has been done, and though only on a limited number of species the results which are available suggest that the range is somewhat similar to our own. Some of these figures are given in Table 7.

Some workers have implied that echo-location similar to that used by bats has evolved in the oilbird *Steatornis caripensis* and the Himalayan or cave swiftlet *Callocalia brevirostris* but others make the point that the birds are using audible sounds in order to locate and here is yet another point of ornithological controversy.

Many writers have remarked on the ability of owls in general, but the barn owl *Tyto alba* in particular, to catch rodents successfully in complete darkness. Some suggested that the bird could be using infra-red energy to locate its prey, but the work of Roger Payne has shown that sound is the clue used by the owl. In a series of ingeniously designed experiments at Cornell University, New York, Payne unravelled the answer. No animal can use a conventional eye in the total absence of light, and he found that barn owls could strike at a rodent to within an accuracy of 1°, but if frequencies above 5Kh were filtered out the bird refused to attack. Infra-red photography has confirmed that an attack in darkness involves moving the head from side to side thus enabling a three dimensional 'audio-pattern' to be built up in a similar manner to a visual parallax.

SMELL

That birds, except the kiwi, have no sense of smell is a statement often made in bird books and in conversations between birdwatchers but whilst olfaction is not well-developed in birds the above statement is far too sweeping. Some responsibility for this error of judgement must be laid at the feet of none other than Audubon who said that whilst vultures would approach and try to eat a picture of a dead animal they ignored its carcase if it was concealed. In 1964 Betsy Bang pointed out that almost all the investigations on the sense of smell in birds had been done either on species with little sense of smell or using materials of more significance to the experimenter than to the species under test. In the same year Stager showed that the North American turkey vulture *Cathartes aura* does much of its hunting by smell, and subsequent work has shown that they gather around areas where underground fuel pipes have fractured and can save the engineers who are looking for the leaks many hours of frustrating searching. Dissection of the brain of the turkey vulture reveals enlarged olfactory organs, and based on the discovery of similar sized organs in other species we may find that smell may play a greater part in the lives of many birds than we imagined. Large olfactory regions have been noted in the brains of most waterbirds, waders, tube-nosed Procellariiformes, nightjars and swifts, as well as the kiwi.

TASTE
Mammals have many taste buds, each being a sort of receptor for detecting one of four so called primary taste sensations; salt, sour, bitter and sweet. This sense is difficult to investigate even in the human species and the only really significant criterion we have to go on is the number of taste buds present. Based upon this, taste does not seem to play such a vital role in the lives of birds as it does in mammals. The rabbit *Oryctolagus cuniculus* on average has 16,500 taste buds, the average passerine bird between 30 and 70 and Allens hummingbird *Salaphorus sasin* only has one! As a rule any taste buds which birds do possess are located mainly in the regions of the pharynx and epiglottis rather than on the tongue, and so it seems that these areas are important in deciding what is finally accepted as food.

We have considered how the bird builds up a picture of its environment by the use of the main sense organs, but also present are nerve cells responsible for detecting such things as pain, pressure, temperature, texture, vibrations, hunger, thirst and proprioceptors situated in muscles and tendons provide a sort of positional sense. None of these has so far been fully explained and much painstaking research is still going on.

## Hormonal or Endrocrine System

Endocrine glands do not have any ducts, and secrete their products, called hormones, directly into the blood stream which transports them around the body. The difference between nervous control by hormones and that by the central nervous system can be explained using a very simple analogy. Nervous control is like a telephone call by which, via the exchange, one

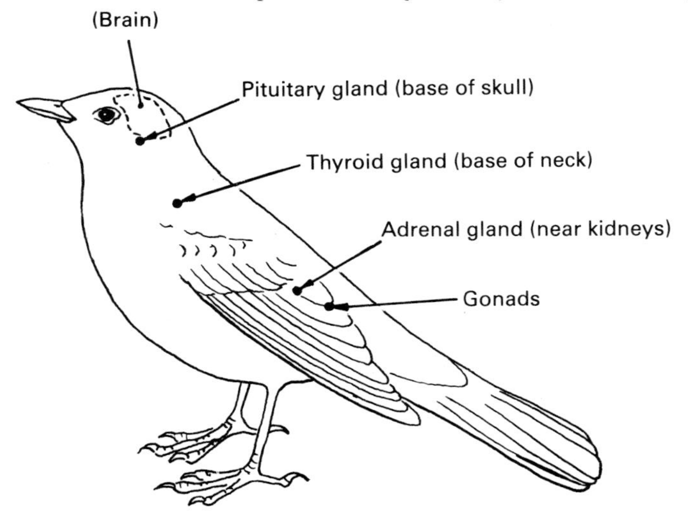


**Fig 86** The relative positions of the four main endocrine glands

individual can be put in contact with another and the message acted upon. The hormonal system, however, functions more like a radio broadcast which can reach many areas, often widely separated. Those not affected by the message can ignore it, and those involved can act accordingly.

One master gland is essential if all these endocrine signals are to be co-ordinated and the 'conductor' of this particular orchestra is the pituitary gland which is situated between the brain and the roof of the mouth. The significance of this and the other main hormone-producing organs and their functions are summarised in Table 8. This system operating in conjunction with the central nervous system and sense organs means that a bird cannot only detect and react to changes in its immediate environment, but can also prepare for long term changes such as migration and bringing itself into peak condition for the start of the breeding season.

Table 8
SUMMARISING THE MAIN ENDOCRINE GLANDS

| Organ | Situated | Appearance | Summary of function |
|---|---|---|---|
| Pituitary | Brain | Made up of two lobes, the anterior and posterior | Anterior lobe produces hormones which control all the other body hormones. Posterior lobe produces one hormone acting on the kidney to conserve water and another to contract the oviduct during egg laying. |
| Thyroid | Base of the neck | Two dark red lobes | Produces hormone called thyroxine which controls growth in most of its aspects; feather development and colour, moult, rate of growth, pigmentation. It may also play some part in migration. |
| Adrenals | Paired glands in front of kidneys | In some species may be fused into the kidney mass. Not divided into cortex | Concerned with production of steroids which regulate blood sugar levels, conversion of proteins to sugar and help activate the sex hormones. Also important in helping the bird adjust to stress. Also produces adrenalin which increases metabolic rate at times when extra efficiency is required. |
| Gonads | Towards rear of body. Close to kidneys | Paired organs | Male hormone is called testosterone, the female called oestrogen. These work in close conjunction with hormones from the pituitary and adrenals. |

*Chapter Eight*

# The Breeding Cycle

Sound plays such an important part in the lives of birds that it is hardly surprising to find that they are splendidly equipped to produce a variety of noises. It may well turn out that this language is much more expressive than hitherto imagined. I have classified these sounds in Table 9, and although nowhere nearly complete it does at least provide a framework on which those interested in bird sounds can build their knowledge. Birdsong as such is found only in the more highly evolved families of the passerine order, but each and every species including the so-called mute swan *Cygnus olor* has its own repertoire of sounds. Let us see just how birds sing before going on to consider what the songs sound like and what their significance is within the whole context of breeding behaviour.

Table 9
SHOWING THE VARIETY OF MECHANICAL AND VOCAL SOUNDS MADE BY BIRDS

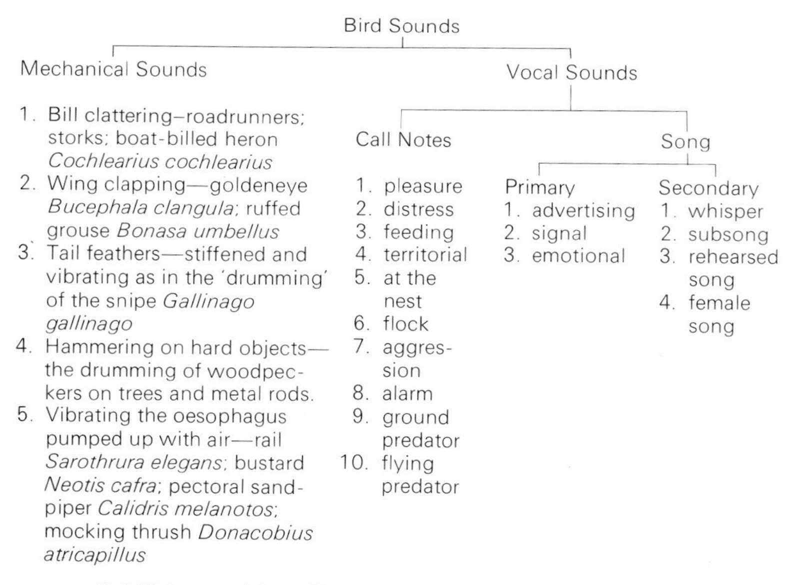


*Ref. Tinbergen, Lister, Thorpe, Nice, Chamberlain and Cornwell.*

## How Birds Sing

We produce our sounds by driving air through the voice box, colloquially called the Adam's apple but scientifically referred to as the larynx. In the

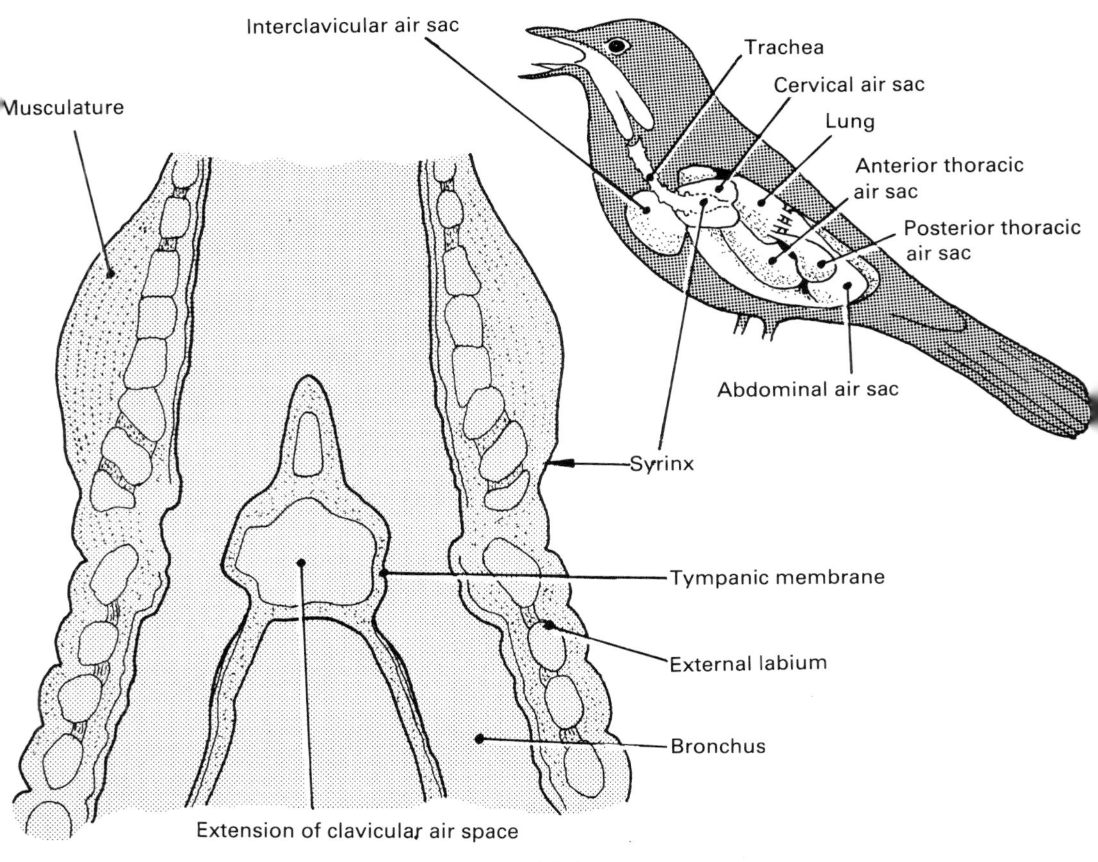


**Fig 87** The vocal apparatus of a typical songbird

bird the larynx is far too near the mouth to be capable of generating the volume of sound required, and it is replaced in function by the syrinx. This shows a great deal of variation from species to species, but three basic types can be recognised. Bronchial syrinxes are formed by several rings, the base of each bronchus having been modified thus producing two structures. This type of syrinx is found in cuckoos, many of the owls and the Eurasian nightjars. A tracheal syrinx is situated at the base of the trachea and is therefore a single structure, an arrangement found in quite a number of the New World passerines including antbirds and ovenbirds. The third type is virtually a combination of the first two, and this rather complex arrangement called a tracheobronchial syrinx is found in most birds (*see* Fig. 87). Its physical structure throws little light on the function, and until the work of Crawford Greenewalt in 1969 there were few satisfactory explanations of how the sound was actually produced, and what follows is largely a summary of his work (*see* Bibliography).

The parts of the syrinx initially involved are the tympanic membrane and its muscles, the external labia and the air sacs. Air is forced into the membrane and pushes it into the bronchial passage. As there are two elements to the syrinx the whole mechanism is doubled, each membrane bulging into its own bronchus. These are controlled by separate muscles and act independently to make it quite possible for a bird to sing a duet with itself. When a bird decides to sing (notice that modern biological thought allows the bird to decide this rather than merely being switched on by instinct) it closes the gap in the bronchus, thus physically separating the lung from the syrinx. This is done by contracting the muscles of the chest cavity causing a negative pressure in the clavicular air space which then bulges into the bronchus. Pressure now builds up in the air sacs as the chest muscles continue to contract, and all the bird has to do is to release either or both of the bronchial closures and air streams over the tensed membranes. Here then is song, loud, controlled and, if required, in duet.

Once the sound has been produced it may well be further modified on the way through the trachea and buccal cavity and in many families the trachea is coiled, making a very effective wind instrument. This situation is typical of cranes and especially the swans.

## Phylogeny and Ontogeny of Bird Song

Phylogeny concerns itself with how a form of behaviour, in this case song, developed in the first place, and ontogeny is defined as the stages through which the song of an individual bird passes as it develops the rhythms typical of its species. Many workers have suggested that in amphibia the voice is used purely as a sex call, and they suggest that this function was the origin of song. Others including Charles Darwin and Eliot Howard indicated that birdsong developed from stress and alarm calls which gradually became strung together. Another point of argument centres around whether the song is genetically determined or whether it is learned. The answer, shown by an analysis of birdsong, would seem to be a compromise between the two.

## Analysis of Bird Sound

One of the things which has always amused me about being in the company of fellow ornithologists is listening to their various attempts to imitate the sounds made by birds. 'Chip-chip-chip; tell tell tell; cherry erry erry; tissy chee-wee-oo.' This is not the name of a new Chinese dish but the sound we think a male chaffinch *Fringilla coelebs* produces in defence of his territory. 'A little-bit-of-bread-and-no-cheese' sings the male yellowhammer *Emberiza citrinella* when all he means is 'all other male yellowhammers keep off my patch'. A look at the descriptions of birdsong given in the old handbooks, followed by a glance at the equivalent section of a modern volume will show how far we have advanced in recent years with regard to the study of song. Gone are the wordy 'chee-ee-ees' and the 'wozzle wozzle wees' and in their place we have oscillograms, or as they are often called sonograms or sound spectrograms, a technique initiated by W.H. Thorpe (*see* Fig. 88). The method involves taping the song and playing it into very sophisticated electronic equipment which analyses the sound in the form of a picture which can be photographed. Initially the method showed that the song of different species produced different patterns, but as the technique improved it has been used to show the stages through which young song-birds pass as they learn to sing, and has shown that the basic song pattern is to some extent instinctive, subsequent embellishments being learned.

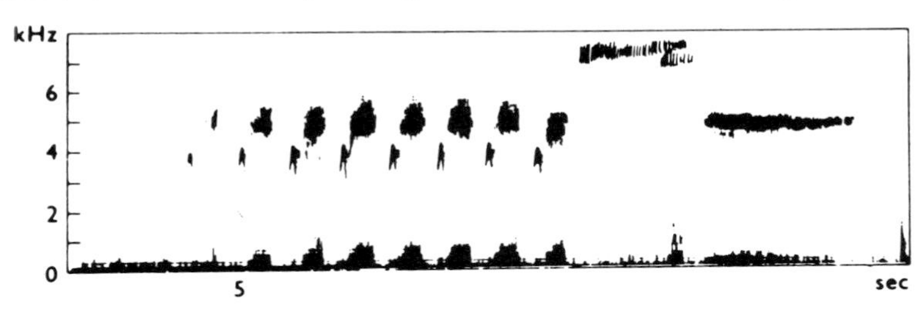


**Fig 88** Sonogram of the yellowhammer *Emberiza citrinella* song, commonly rendered as 'a-little-bit-of-bread-and-no-cheese'

The onset of the bird's first breeding season seems to be the period when the basic song is really modified, no doubt triggered by a liberal dose of hormones. Painstaking work on the chaffinch has shown that there are song variations apparent in different parts of the range—in other words, birds have dialects, just as we do. Further work will doubtless enable individual birds to be recognised.

## The Meaning of Bird Song

> I could not sleep again, for such wild cries,
> And went out early into their green world:
> And then I saw what set their little tongues
> To scream for joy—they saw the east in Gold.

W.H. Davies is here describing the dawn chorus, and he is sure that the birds are enjoying life, an opinion which later appalled many behavioural scientists. Modern workers believe that song is mainly a functional thing, but there can surely be no doubt that many birds enjoy being perched high on a branch singing to their mates, or calling to rivals from deep within a bush informing them that the ideal nesting spot is already taken. Some have suggested two components which they have labelled a signal song, which co-ordinates the mated pair, and an emotional song, by which the bird gets rid of excess energy. For those who allow themselves to become anthropomorphic, this is the element which the bird enjoys and it is in this component where individual variations can be given full expression.

## Initiation of the Breeding Season

It seems quite certain that secretions from the hypothalamus set in motion the flow of hormones controlling the growth of the testes and ovaries, but we begin to find difficulties when we try to discover what stimulates the hypothalamus itself. Most folk living in north temperate regions would say that the trigger is the coming of spring when the days begin to get significantly longer. Photoperiodism, as this is called, doubtless plays its part in bringing many species, including the chaffinch, into breeding condition, but doubtless there are other factors also at work.

Some birds, including the Hawaiian goose, otherwise known as the ne-ne, *Branta sandvicensis* actually nest as the photoperiod is shortening as do lyrebirds (*Menura* spp.) and the emus (*Dromaius novaehollandiae*). This means of course that they breed during the winter. The Emperor penguin *Aptenodytes forsteri* lays its egg in mid July when the Antarctic is facing an horrific winter in total darkness.

Other environmental factors have been shown to have their effect on the initiation of breeding, including both rising and falling temperatures, and the sight of green grass following rain stimulates the African red-bellied weaver *Malimbus erythrogaster* to commence its breeding ritual. The onset of rainy weather in areas where this occurs at infrequent intervals may actually cause certain species to breed twice, in an effort to raise the population whilst the going is good. This is certainly the case with a lot of Australian birds which have been reported to breed twice within a six month period following a period of unaccustomed rain, a fact also noted regarding Abert's towhee *Pipilo aberti*. The whole timing of the egg laying must, of course, be geared to the young growing up at a time when their food supply is at a maximum. Whatever combination of factors triggers the development of the gonads, the hormones which they produce bring the birds to a frenzied peak when courtship and pair formation become their sole preoccupation. Before breeding can begin, however, some spot, free from inter and intra-specific interference, must be set up where eggs can be laid and incubated and the young raised. This is the origin of territorial behaviour.

## Territory

The human species has been interested in the study of birds ever since it was discovered that they were good to eat, and learned works have been published on ornithological topics since Aristotle. Although migration was frequently mentioned, it took until the publication of *Der Vogel und sein Leben* written by Altum in 1868 for territory to get a mention, and it was not until Eliot Howard published his *Territory in Bird Life* in the 1920s that attention began to be focused on this matter. The picture is still somewhat blurred, however, and the concept is so variable that it is difficult to define.

In 1941 Nice suggested seven types of territory but this was simplified somewhat by Hinde in 1956 who listed four types.

TYPE 1

A large breeding territory inside which the whole spectrum of breeding activity is enacted. This is typical of most passerine species particularly the Old World warblers such as the chiffchaff *Phylloscopus collybita* and the willow warbler *Phylloscopus trochilus*. Incidentally these two species look so much alike that they are best distinguished by their very different songs. This type of territory is also typical of woodpeckers, cardinals and mockingbirds.

TYPE 2

A large breeding area which provides everything required by a breeding pair of birds except an adequate food supply. Such a territory is typical of many grebes, swans, harriers, the oystercatcher *Haematopus ostralegus*, and the fieldfare *Turdus pilaris*.

TYPE 3

This is defined as a quite small territorial area used purely to site the nest, and the only area actively defended is the small space around this area. Some non-colonial birds such as pigeons favour this type of territory but it is typical of the colonial nesters such as the social weaver *Philetairus socius*, the rook *Corvus frugilegus*, the gulls, the grey heron *Ardea cineria*, the shag *Phalacrocorax aristotelis* and the brown pelican *Pelecanus occidentalis*.

TYPE 4

Here the territory is merely an area set aside for a sort of ritual copulation and not for nesting. These are often referred to as leks (from the Norse word for 'to play') or display areas. This type of behaviour is found in ruffs *Philomachus pugnax*, the black grouse *Lyrurus tetrix*, the bowerbirds Ptilonorhynchidae, birds of paradise Paradisaeidae, and the capercaillie *Tetrao urogallus*.

Whilst territory formation is undoubtedly the rule, there are exceptions to every rule and birds such as the American redshank *Totanus totanus* and the scarlet-rumped tanager *Ramphocelus passerinii* are just two examples of birds which form no territory. However, since the majority of birds do, at sometime or another, establish a territory we must assume that it bestows an evolutionary advantage, a point underlined by the ferocity with which they are often defended.

## Defence of Territory

Any bird trespassing on the territory of another of its species is immediately greeted by a threat display, either by posture or, in passerines, by song and/or posture, and, if this fails, by physical attack although the rules may be bent to allow neighbours to visit a drinking pond or stream. Threat and attack were investigated by David Lack in his classic work on the life of the European robin *Erithacus rubecula*, and he discovered that the breast is the trigger for attack: hide this and the rival is tolerated; a stuffed specimen with a red breast is not. In 1972 Smith found that if he blacked out the colours on the red-winged blackbird *Agelaius phoenicus* 64 per cent lost their territory, but if rival males were artificially restricted from the territories the discoloured birds could attract a female and mate with her. It is often noted in these days of automobiles with reflective hub caps that birds often attack their own reflections thinking them to be rival birds intruding into their territory. Lack's work on the robin also showed that male birds without territory seemed to suffer from a depressed sexual drive and were unable to mate. It is thus inferred that the possession of the correct type of territory for the species brings on the next stage of the breeding cycle which is courtship.

## Courtship and Display

One of the sexes, almost always the male, is ready to copulate before the other, and it is at this time that courtship plays a vital role in bringing the partner to a point where mating is inevitable. There seems little doubt that this is the origin of the very pronounced sexual dimorphism found in some species. Sexual displays are not a waste of surplus energy, but are vital to the survival of many species, and it was Lorenz who pointed out that displays function as releasers, one ritualised movement following another and playing its part in the build-up to the climax of the breeding cycle. These complex displays are unique to the species and do much to keep each discrete and they are of great interest to the ornithologist. What better sight is there in the natural world than the nuptual dance of the great crested grebe *Podiceps cristatus* which is a highly organised ritual of one reflex action triggering the next. In the course of time the most effective males will be those who are the best adorned, and some exaggerated patterns have evolved, including the fanned tail display of the male peacock *Pavo cristatus*, the birds of paradise, Paradisaeidae, and of course the great variations found in the ruff *Philomachus pugnax*. This gradual exaggeration of sexual characteristics is called hypertely.

In 1954 Niko Tinbergen recognised four separate aspects to courtship display including acting as what he called biological isolating mechanisms. He also stated that display tended to concentrate the attentions of the mate on breeding to the exclusion of all else. Thirdly he pointed out that the whole object of display was to attract a potential mate into the territory and keep its attention and interest long enough for mating to occur. Finally in the case of group displays particularly among seabirds, this mutual display

**Fig 89** Chick exchange display of the king penguin *Aptenodytes patagonicus*

stimulates not only the mate but the whole colony, which is eventually brought to a frenzy of mating thus bringing peak laying and hatching of young which will then have the maximum protection possible since all the adults will be keen to defend the colony. Once a male and female begin to show an interest in each other they are said to have developed a pair bond, and display, copulation and bringing up a family may strengthen this pair bond, though its actual duration depends upon the species concerned.

## Duration of the Pair Bond

Some species such as swans, geese, the common terns *Sterna hirundo*, the raven *Corvus corax*, the Carolina chickadee *Parus carolinensis* and the wren tit *Chamaea fuscica* pair for life and the divorce rate is sufficiently low for the pair bond to be described as permanent. Birds such as house martins *Delichon urbica* and the swallow *Hirundo rustica* are faithful to a nest site rather than to a particular mate. The males tend to return from migration first and set up territory; if the female also happened to survive the rigours of the return trip the pair bond of the previous year may well be re-established. In many ducks the sexes may form a pair bond at the beginning of winter, and may then migrate together; it seems that it is the male who follows the female and he will be someone she picked up on the way to her breeding grounds. When all her eggs have been laid and incubation starts the wandering suitor deserts his spouse and the pair bond is broken. This type of behaviour is typical of the mallard *Anas platyrhynchos* and the goldeneye *Bucephala clangula*. Even shorter is the pair bond formed in birds which lek such as the ruff *Philomachus pugnax* and the black grouse *Lyrurus tetrix*.

## Egg Laying

Following display, successful reproduction depends upon both sperm and eggs being fertile. The sex organs or gonads (*see* Fig. 86) are situated just in front of the kidneys: in the male there are a pair of testes which are usually oval but one is often larger than the other; in the female this process has gone one step further and only the left ovary remains. During mating the birds press their cloacae together and the sperm is passed into the female. The usual mating position is for the male to balance on the female's back, but a sort of cloacal penis is possessed by the ratites, ducks and a few of the gallinaceous birds. The end result of this is a fertilised egg which is then laid.

Some birds, the herring gull *Larus argentatus* being a good example, are said to be determinate layers and once they begin to lay produce a definite number of eggs to make up a clutch (an average of 3 in this case) regardless of external factors; in other words the clutch size seems to be genetically determined. Other birds, including the domestic fowl, are known as indeterminate layers and carry on producing eggs until a given number are physically present in the nest. The eastern flicker *Colaptes auratus* normally lays up to 7 eggs in a clutch, but if, early in the laying sequence, an egg is carefully removed every time a fresh one is laid the bird goes on and on and can be induced to lay up to 70 eggs.

The different breeds of domestic hen vary a little in their behaviour, but commonly a hen begins by laying an egg soon after dawn. Each succeeding day an egg is laid approximately an hour later than the day before, and this continues until the laying time falls late in the afternoon when a day is skipped; the next egg appears on schedule early the following morning, however, and then the sequence is repeated. The maximum number of eggs seems to be produced during the first year of its life, and then under comparable conditions with regard to food, temperature, daylight and careful husbandry the number of eggs the hen lays is roughly 75 per cent of its yield over the preceding year.

In the wild there is a similarity between litter size in mammals and clutch size in birds, and the basic rule is that the more young which are produced then the more expendable they are—a harsh fact of life. Conversely, the fewer young there are, the more parental care is lavished upon them, for the continued success of a species means that the number of newly born individuals surviving to breed must be at least equal to those which die. Those birds which have few natural enemies coupled with a long life expectancy will produce fewer eggs than those with many natural enemies and a high mortality rate.

## The Avian Egg

### STRUCTURE

An avian egg may be described as a fertilised ovum wrapped up snugly inside a case which both protects and feeds it. Once laid, all subsequent development within it depends upon material deposited there prior to laying (*see* Fig. 90). The only exception to this is the entry and exit of respiratory gases through the permeable calcareous shell. The ovum passes down the genital tract pushed along by involuntary muscle movements (called peristalsis) before entering the shell gland which it spirals around for the protective shell to be layered on to form the egg. The actual thickness of the shell varies quite a lot, being quite thin in the woodpigeon *Columba palumbus* but much thicker in some birds, and in the francolins almost 30 per cent of the egg weight is shell. It is now well known that certain pollutants affect the shell deposition especially in predatory birds such as the peregrine *Falco peregrinus* and the sparrowhawk *Accipiter nisus*. These thin-shelled eggs are too fragile to be successfully incubated and breeding fails because of it; this is such an important aspect of present day bird biology that we shall return to it at length in Chapter 12.

The quantity of yolk within an egg varies a great deal from species to species and in some ducks it can be up to 50 per cent whilst in some cormorants this figure can be reduced to less than 15 per cent. A look at an unfertilised egg will reveal an opaque circular white spot on the yolk. This is called the blastodisc, and it and the much paler white yolk around it are less dense than the yolk and float on top of it. Within this blastodisc is the germinal vesicle which is the female reproductive cell. If this has been fertilised by sperm cells swimming up from the cloaca and fusing with the ovum prior to shell deposition then it commences to divide, initially to

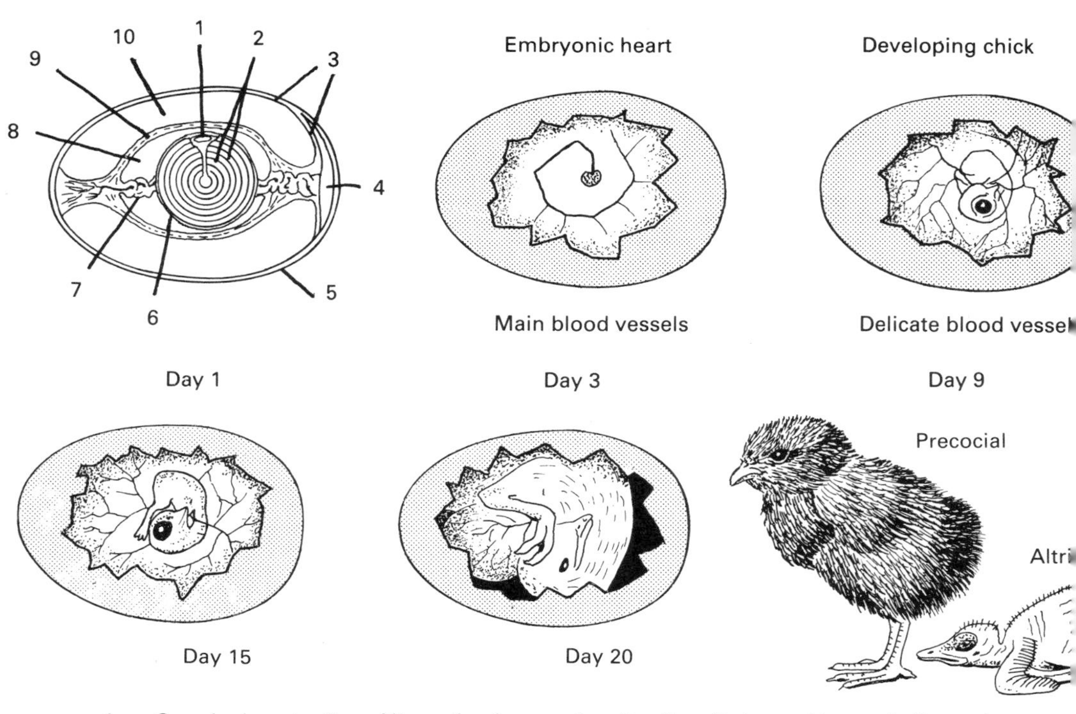


1. Germinal spot 2. Alternating layers of yolk 3. Outer and inner shell membrane
4. Air cell 5. Shell 6. Membrane enclosing yolk 7. Chalaza
8. Inner layer of albumen 9. Fibrous layer 10. Outer layer of albumen

**Fig 90** Egg development of the domestic hen. The chick is precocial and is shown alongside an altricial chick which will need to be looked after by its parents for many days

produce the germinal disc or blastoderm, and eventually to form the chick. The yolk of an egg is laid down in concentric layers, white layers being easily distinguished from the yellow yolk which contains more fat and pigments. Surrounding the yolk is the white albumen, and although it does not seem so when looked at superficially this is also laid down in an organised manner consisting of four concentric layers. The inner ring is called the chalaziferous layer and part of it forms the chalaza which is a sort of system of 'supporting ropes' which tend to hold the yolk and stabilise its position. When the egg is turned the chalaza twists and turns, holding the yolk steady. Then comes the inner liquid layer in which both yolk and chalaziferous layer float. Next comes the albuminous sac which is viscous, and it is to this structure that the chalazae are anchored and which cushion and protect the yolk. The area is attached to the shell at either end by fibrous structures, each called a ligamentum albuminis. Finally we have the outer liquid layer which supports the denser structures thus preventing damage during movements of the egg, and also cushioning the growing chick.

**Fig 91** The five basic avian egg shapes

EGG SHAPE AND COLOUR (*see* Fig. 91)
The original colour of eggs was probably white since the reptilian method is to bury them in sand or loose soil; this pale colour is retained by primitive families such as cormorants, pelicans and albatrosses and by more advanced forms which nest in holes. This means that they have never needed to develop cryptically coloured eggs, and indeed those of owls and kingfishers have remained white because they have always laid eggs away from daylight. Those nesting in holes, such as certain titmice which lay speckled eggs, may have taken up this habit after a period of nesting out in the open.

Most pigments are incorporated into the superficial layers of the shell at the time this structure is produced in the shell gland. They are derived by the breakdown of blood cells in the genital tract. They are either cyanins, which come from the bile pigments and which give green and blue colours, or they are red, brown and black colours which are derived from the breakdown of haemoglobin. The value of colouration in eggs is of course in the camouflage which they provide, and in many ground-nesting birds such as some waders they are of great survival value.

## The Nest

The megapodes are the only family in which no form of incubation by the birds themselves is carried out, the eggs being buried in vegetation or sand where the decaying material or heat of the sun powers the development. This is close to the original method used by the bird's reptilian ancestor, but as birds are warm-blooded creatures the eggs, if the chicks within are to develop properly, need to be kept at relatively high temperatures. It is in response to this requirement that nest-building evolved. Some birds' nests are very simple, almost haphazard, whilst with the more advanced nest-builders the whole process involves a great many complex interlinked movements, many of which are unique to the species and bespeak many years of evolved specialisation.

Many of these nesting patterns have enabled some species to breed in an area rich in food, but which could not have been exploited without a nest to cradle the eggs out of reach of most predators. Many birds including the European chaffinch *Fringilla coelebs* deliberately add lichens to the nest which match the bark of the tree in which it is placed, whilst others such as the magpie *Pica pica* build domed nests which bristle with thorns; the kingfisher *Alcedo atthis* and Manx shearwater *Puffinus puffinus* are examples of birds nesting in burrows, whilst the goldcrest *Regulus regulus* slings its nest like a basket beneath a branch often too delicate to support the weight of predators, and thus effectively insulates the species, an idea even more fully exploited by the weaver birds (*see* Plate XI, p. 144).

**Fig 92** An eider duck *Somateria mollissima* incubating her eggs

INCUBATION

This may be undertaken by the female alone as in the domestic fowl, by both parents as in the majority of species, or in a few cases such as the phalaropes by the males alone. Broodiness is defined as a psychophysiological state controlled by delicate hormonal balance. During incubation special patches of down feathers on the belly moult to reveal the skin beneath, which is particularly rich in blood vessels. These warm 'brood patches' are pressed close against the eggs. Incubation is normally delayed until the clutch is complete so that all the young hatch together, but in some birds, the barn owl *Tyto alba* for example, incubation begins after the laying of the first egg. This means that the eggs laid first will hatch first and in times of food shortage the eldest young eat their smaller and weaker brethren. This sounds bizarre in human terms, but it is surely important to the survival of a species that a few individuals should survive at the expense of the others rather than all die of starvation.

## The Chicks

Some birds produce precocial chicks which hatch completely covered in down feathers, are able to follow their parents almost immediately, and can cope with feeding themselves even though the parents may have to locate the

**Fig 93** The egg tooth can be clearly seen. It is shed soon after hatching

food for them. The passerine birds like thrushes and warblers hatch helpless and almost naked, a condition referred to as altricial (from the Latin 'altrix', meaning nurse) and they do require prolonged parental care. Altricial birds typically build well-constructed nests and lay few eggs, but infant mortality is much lower and so the net result of both methods is more or less the same. The older terms for precocial and altricial were nidifugous, which literally means 'fleeing the nest', and nidicolous, 'nest-loving'.

Prior to hatching the chicks begin to 'cheep' and are thus in contact with each other and with their parents; it has been suggested that this contact may help to ensure that all the chicks hatch at approximately the same time. When ready to hatch the chick possesses two structures which assist in its escape from the confines of the shell. On the top of the upper mandible is a calcareous point called the egg tooth (*see* Fig. 93), and there is also a hatching muscle which enables the body to be flexed thus using the egg tooth like a hammer—both structures disappear soon after hatching.

## Feeding the Chicks

The mode of feeding depends upon whether the young are precocial or altricial, and which parent actually attends to the feeding depends upon the type of pair bond linking them. The male passerine is usually much too concerned with protecting the territory during the period of nest-building, laying and incubating to be involved with care at the nest, but once the chicks have hatched he seems confident that he has successfully excluded rival birds of his own species and helps in the feeding of his offspring. In many birds a second and even a third brood is produced and in these cases the assistance of the male is absolutely vital. In the coot *Fulica atra* the young are precocial, but they do not all hatch at the same time and whilst the female takes some of the young in search of food the male completes the incubation. In some species like the moorhen *Gallinula chloropus* the young of one brood may help to look after succeeding broods, and the first clutch of swallows *Hirundo rustica* and house martins *Delichon urbica* are also reported to assist in the feeding of subsequent broods. There are also occasions when unmated birds may join in and help the rightful parents attend to the young. Skutch studied this phenomenon in the Mexican jay *Aphelocoma ultramarina* and it has also been tentatively suggested that other corvids and rails may also practise this form of behaviour; it is however very difficult to prove that this occurs with any degree of regularity and ornithologists are sceptical regarding its true value as an aid to species survival. There is a recent suggestion, however, of cooperative breeding among babblers.

The actual rate of feeding of altricial young varies a great deal; some small birds may feed their young as frequently as 40 times in an hour, and at this rate both parent birds would be working at a frantic pace. Fortunately

only every other year. The actual length of this period will vary depending upon the availability of food and prevailing weather conditions; this is particularly evident in the case of the swift *Apus apus*. In David Lack's extensive study of the swift (*see* Bibliography) he found that in the cold weather when few flying insects were about the young went short of food. In the early stages of development altricial birds are unable to control their body temperature and most must be brooded when the temperature falls, but young swifts have another stratagem. When deprived of food they go into a sort of suspended animation, having some similarities to hibernation, in which heart rate and temperature both fall and thus reduce the amount of food needed. On the return of better conditions the nestlings quickly return to normal. Whilst this is an extreme case, the development of all birds, especially those which are altricial, must be affected to some degree by these factors.

The time of leaving the nest again varies considerably but some young birds such as the skylark *Alauda arvensis*, corn bunting *Emberiza calandra* and short-eared owl *Asio flammeus* depart before they can fly, and skulk about in the bushes still begging food from their parents. The evolutionary advantage of this behaviour would seem to suggest that a nest full of noisy young if predated would result in the loss of the complete clutch whilst if the vulnerable young are spread about, some, at least, will survive. Other species may keep their young in the nest until they can fly; this is the case in birds such as swifts, the swallow *Hirundo rustica* and house martins *Delichon urbica* which are aerial feeders and whose nests tend to be a little more secure from predators.

Before leaving the subject of bird breeding we should take a look at brood parasitism. What on the face of it seems to be very simple turns out to be very complex since the parasitic bird has to select a host then lay an egg capable of fooling the foster parent. The egg must hatch out before those of the other eggs in the nest, thus giving a survival advantage to the chicks of the parasite. There are so many variations employed by the various brood parasites that we are forced to the conclusion that the methods have evolved independently and indeed are still evolving. This is seen in the North American cow birds *Molothrus* spp. and especially in the European cuckoo *Cuculus canorus* where a particular female specialises on one particular host species. Thus we have those parasitising meadow pipits *Anthus pratensis*, others preferring dunnocks *Prunella modularis*, and some favouring reed buntings *Emberiza schoeniclus* or reed warblers *Acrocephalus scirpaceus* (*see* Plate VI, p. 140). It is just feasible that in time natural selection could produce separate species of cuckoo. Looked at through anthropomorphic eyes we often feel a sense of revulsion at the antics of the sly and devious cuckoo, but we must take care not to forget that all is fair in nature—the whole purpose of a species is to remain viable, and any method which achieves this end is acceptable.

*Chapter Nine*

# Migration

The word migration usually makes us think of long hazardous journeys as the wild geese leave the ice-bound winter tundras for the warmer south, whilst in spring swallows and warblers head north across desert wastes to breed in the temperate lands. If pressed to expand upon this definition most people could mention the periodicity of migrations, and can thus distinguish migration from irruptive movements made in response to a particular event and which do not involve a controlled return journey.

This type of irruptive movement sometimes occurs in the case of the snowy owl *Nyctea scandiaca*, normally an Arctic dweller. It feeds mainly on the lemming *Lagurus lagurus*, but the well-documented fluctuations in the populations of these rodents tell us that every four years or so their numbers crash dramatically; the snowy owl, at these times, must starve or move south in an irruptive movement. The same situation is found in the crossbill *Loxia curvirostra* which feeds almost exclusively upon pine cones, and should these be in short supply the birds must travel or perish. Other irruptive

**Fig 95** A snowy owl *Nyctea scandiaca*, an irruptive species in Britain

species include the waxwing *Bombycilla garrulus* whose diet consists of seeds and berries, or the Siberian nutcracker *Nucifraga caryocatactes macrorhynchus* which eats the seeds of the Arolla pine *Pinus cembra*. A related North American species, Clark's nutcracker *Nucifraga columbiana* is also irruptive as is the Asiatic Pallas's sandgrouse *Syrrhaptes paradoxus*.

In true migration the movement is very regular and involves exploiting the best areas for food supply, even if this entails journeying thousands of miles: a bird which feeds in the south and breeds in the Arctic may be exploiting a very rich supply of food, albeit of limited duration. Migration occurs in only about 15 per cent of bird species, but must have survival value to those which favour it. In times of adverse climatic conditions other birds may either put up with them or hibernate.

## Hibernation

This phenomenon has not been studied very deeply in birds, and only one species, the North American Poorwill *Phalaenoptilus nuttallii*, has been regularly found in a torpid condition in the dark cold days of winter, its heart beat, breathing rate and other metabolic processes reduced to a minimum. Numbers are found huddled together in cliff crevices, their body temperatures down from 41°C (106°F) to only about 17°C (64°F). This behaviour has been induced under laboratory conditions and we can safely assume that the species is an habitual hibernator. We have already referred to the torpid condition developed by swifts *Apus apus*, and in 1907 Hughes mentioned young birds surviving a fast lasting 21 days. For the first few days the body temperature seems steady at about 40°C (102°F) but should the air temperature fall below 18°C (66°F) then the swift's temperature rapidly falls to about 15.5°C (50°F). Even the adult birds deprived of flying insects are capable of surviving for a few days in a torpid condition, and Lorenz reported swallows *Hirundo rustica* behaving in a similar way when hit by cold weather during their migration over the Alps. It was reported in the sixteenth century that ptarmigans *Lagopus mutus* dug holes in the snow and stayed there in bad weather. This does in fact occur, but we do not regard this as true physiological hibernation, and hibernation in the bird world is the rare exception rather than the rule.

## The Evolution of Migratory Behaviour

Four possible factors have been suggested as causes for migration but none is totally acceptable in isolation and there is no doubt that a combination of these and other factors is involved. Listed in order of popularity we have: the effects of the Ice Ages; a return to some hypothetical ancestral home; cold weather; and food shortages, though this last is least tenable as most species leave their summering grounds long before there is any food shortage.

By far the most publicised theory has been the north- and southwards movements of the ice floes during the Pleistocene period (*see* Table 1, p. 14) and some workers, particularly in Europe where the effects are rather more obvious than in other regions, have intimated that this is the only factor needed to account for migrations. More recent opinion shows that this ice-sheet theory, although important, has been over estimated, since the Ice Ages occupy only a fraction of the time scale over which migratory movements have been evolving.

The theory of a return to some ancestral home is based upon Wegener's continental drift theory which postulates the presence of two original land masses which fractured into the shapes of the present continents which have since gradually drifted apart. Birds which moved between one area and the next on the original land mass are now obliged to make the journey across a gradually widening width of water. It should be noted, however, that this fracturing and continental drifting was already well advanced when birds first evolved, but the theory has some bearing on our understanding of bird movements. The ancestral home can often be worked out by assuming that birds of tropical origins will tend to arrive in the breeding grounds late and depart for their winter quarters early. In Europe the cuckoo *Cuculus canorus* does not arrive in its breeding grounds until late April and departs for Africa during August, a pattern followed by the swift *Apus apus*. In North America the barn swallow *Hirundo rustica* shows a similar behaviour pattern. It would seem that many birds do actually follow the ancestral routes which developed following the recession of the ice-packs, and when temperatures were a little warmer than at present. This type of behaviour explains the migrations of European birds such as one form of the wheatear, *Oenanthe oenanthe leucorrhoa*, which now penetrates as far north as Greenland from its wintering grounds in tropical Africa. Wolfson applied Wegener's drift theory to account for the migrations of several wading birds such as the sanderling *Calidris alba* which reaches the highest latitudes to breed and winters in tropical areas, and the knot *Calidris canutus* which can journey so far south in winter that it may reach Australia, New Zealand, Patagonia, India and South Africa all now widely separated from its breeding grounds in the high northern latitudes. According to the theory, all these areas once formed part of the huge land mass called Gondwana which eventually broke up and destroyed the ancestral home. Whilst superficially attractive, the theory does not stand up to really close scrutiny and must remain as a part of a complex solution rather than constituting the whole answer.

Weather, both long and short term, also affects migration patterns. Our own life span is far too short to enable us to appreciate long-term weather changes and this is why the diaries of naturalists can prove of great value by recording faithfully all events, however trivial they may seem at the time. The climate of Europe, and especially Britain, seems to have ameliorated in recent years; gone are the winters of extreme cold and scorching hot

**Fig 96** Hirundines gathering on migration, a common sight in temperate climates in the autumn

summers—one seems to have merged into the other. Should we eventually return to more extreme conditions, new migratory patterns could well be produced. The collared dove *Streptopelia decaocto* for example began to extend its range from southern Europe into Britain during the 1950s and has now reached pest proportions in some areas. Should regular cold winters once more become the rule then the collared dove will either become extinct in Britain or develop into a migrant, moving south in the winter and north in spring the penetrating distance being dependent upon extremes of tolerable temperature.

We have seen enough of the origins of migration to appreciate its complexity, and a complete explanation defies even the most sophisticated modern techniques. The ancient world knew something about bird movements, and it was many years ago that the author of the Bible's Book of Job asked,

> Is it by Thy wisdom that the Hawk soareth
> And stretcheth her wings towards the South?

Indeed the history of our ideas concerning migration goes back to times prior to written history since *Homo sapiens* the hunter simply had to be aware of the movements of game or he would have soon become restricted to very few regions—or become totally extinct.

## Historical Ideas on Migration

Early writers recognised as clearly as we do today that birds which were unable to cope with prevailing conditions either migrated or hibernated; the only difference between ancient and modern thought concerns the relative importance and frequency of the two strategems. Aristotle reported swallows *Hirundo rustica* being found naked and torpid on the sunny sides of mountains, and he also reported the turtle dove *Streptopelia turtur* as moulting its feathers and hibernating. In 1485 we find von Caub in his *Ortus Sanitatis* stating that swallows hibernate, thus following, as dictated by the prevailing religious doctrine of the time, the writings of the ancient world. Von Caub did, however, give an accurate description of the migration of the stork *Ciconia ciconia*. Gesner published a monumental five-volume work in which he repeats many older theories, but he does come up with some clear descriptions which are not out of place among our most modern thoughts:

> Some birds are resident, such as doves; others like swallows stay only six months. We have watched them arrive in March and leave in August. Swallows go to more temperate places for the winter if they are nearby; but if they are distant they hide in the localities where they are.

At least this passage shows that thoughts concerning migration were showing some signs of compromise, but in 1555 Olaus Magnus, the Archbishop of Uppsala pushed the emphasis firmly backwards into the arms of Ancient Doctrine:

> Several authors who have written at length about the inestimable facts of nature have described how swallows often fly from one country to another, travelling to a warm climate for the winter months; but they have not mentioned the denizens of [birds] in northern regions which are often pulled from the water in a large ball. They cling beak to beak, wing to wing, foot to foot, having bound themselves together in the first days of Autumn in order to hide among canes and reeds. It has been observed that when spring comes they return joyously to their old nests or build new ones according to the dictates of nature. Occasionally young fishermen unfamiliar with these birds will bring up a large ball and carry it to a stove where heat dissolves it into swallows. They fly, but only briefly, since they were separated forcibly rather than of their own volition. Old fishermen, who are wiser, put these balls back into the water whenever they find them.

In Europe arguments regarding hibernation versus migration raged on. The great Swede Linnaeus lined up with Barrington the eighteenth century Englishman in favouring the hibernation of swallows, as did Cuvier writing in 1817, but the Frenchman Buffon kept swallows in an ice-box and found that they died, and the anatomist John Hunter (*see* Chapter 5) pointed out that lungs and air sacs were ill-adapted for prolonged immersion in water, and gradually opinion shifted in favour of migration. The New World, however, supported new ideas much more readily, and Oviedo, recording the movements of predatory birds between 1526 and 1535, and noting the often vast populations wrote,

> I have seen them pass over Darien and Nombre de Dias and Panama ... and this passage ... continueth a moneth or more about the moneth of March.

Although Oviedo clearly accepted bird movements, he assumed, because he did not notice any return flight, that the birds continued around the world. Willughby in 1678 was vigorously supporting migration and J.J. Audubon wrote voluminous works on the subject and was the first to decide upon a scientific approach involving the physical marking of birds.

## Studying Migration

In recent years many methods have been developed, with varying degrees of success, in an attempt to make the story of migration a little easier to understand. A few of the less bizarre methods are looked at below.

### MOON WATCHING AND FOG LISTENING

This involves focusing a telescope on the full moon and observing migrating birds passing across its disc. This gives an excellent method of counting numbers but is greatly limited because it is always difficult and often impossible, especially with small birds, to identify the species. Even on dark and foggy nights ornithologists skilled in the recognition of bird sounds can listen to birds passing overhead. A more advanced method is to record the sounds on tape and from these a sonograph can be produced (*see* Chapter 8). This method is, however, never likely to be much more than an interesting diversion, producing on its own very little more than the valuable notes of an amateur naturalist who records arrival dates of visitors and can also supply the last date on which a migrant species was seen.

### COLLISION WITH SOLID OBJECTS

Birds on migration often fly into tall objects such as radio and television masts and especially lighthouses. Ornithologists have enlisted the help of those who staff these prominent structures and who collect, identify and count the nightly casualties. Thus the peak periods of migration of many species have been worked out. It should be pointed out that some effort has been devoted to making lighthouses safer for the birds. Keepers living in these often isolated areas have proved keen to help research of this nature, pointing out that it gives them something different to do, and I know several keen and competent ornithologists who began their studies in just this way.

### EXPERIMENTS ON CAPTIVE BIRDS

It has been found that if a bird is held captive it becomes restless round about migration time and even if it has insufficient space for flight it will hop or walk in the appropriate direction. These movements have been recorded using very sophisticated electronics. Other birds have had lights attached to their feet so that their initial flight movements following release could be followed and an even more sophisticated (and expensive) technique has involved fitting birds with radio transmitters and following them in a light aircraft. Another method has been to transport birds varying distances from the nest site and time their return, a subject to which I will return later in this chapter.

### RADAR

During World War II research into 'early warning systems' for detecting enemy raiders developed quickly. Problems were encountered early in the development because in Britain large seabirds, in particular the gannet *Sula bassana* were capable of producing a blip of sufficient size to panic the inexperienced observers: this stimulated the boffins to make refinements. By the early 1950s the equipment was becoming more and more sophisticated, but it still suffered from what were assumed to be meteorological interferences at certain times of the year particularly autumn and spring. These, for want of a better description, were called 'angels'. Two pieces of evidence put forward independently in 1957 by Harper in Britain and Sutter in Switzerland proved that the 'angels' were congregations of small birds. Now ornithologists had a tool which would tell them the altitude of migrating birds and also their speed of flight, both sets of data difficult to obtain by any other method on a large scale (*see* Table 10).

Table 10
SHOWING FLYING SPEEDS AND ALTITUDES OF SELECTED SPECIES AS DETERMINED FROM RADAR DATA.

| Species | Ground Speed in kph (mph) | Altitude in m (ft) |
|---|---|---|
| *Accipiter nisus*–sparrowhawk | 41 (26) | 860 (2,800) |
| *Hirundo rustica*–swallow | 53 (33) | 800 (2,600) |
| *Sturnus vulgaris*–starling | 69 (43) | 1,080 (3,500) |
| *Corvus frugilegus*–rook | 59 (37) | 3,390 (11,000) |
| *Anas platyrhynchos*–mallard | 85 (53) | 3,230 (10,500) |
| *Cygnus cygnus*–whooper swan | 70 (44) | 3,390 (11,000) |
| *Fringilla coelebs*–chaffinch | 41 (26) | 2,000 (6,500) |

*Note. These figures are approximate and vary with conditions, but they demonstrate the data which can be collected by radar.*

All countries quickly adopted this method and skilled workers can now recognise the patterns produced by large and small passerines, waders, gulls, waterfowl and birds of prey although individual species cannot be identified. If careful and continual records are made, including details of flight patterns and alterations of course and if notes are also made of weather conditions, a great deal of valuable data quickly accumulates. It is possible to work out which birds are affected by wind, rain, fog and so on, and this will obviously throw some light on the very interesting question of how birds navigate.

RINGING (OR BANDING)

As long ago as Roman times adult swallows were removed from their nests and taken in cases to chariot races. A coloured thread was tied to the bird's leg carrying information about the winner of the race; on release the bird flew straight back to its nest where eager punters were waiting. Early attempts were made to mark birds which belonged to influential individuals, such as the peregrine falcon *Falco peregrinus* belonging to King Henry IV of France (1589–1610) which complete with golden ring escaped from the mews at Fontainebleau and was discovered a day later on Malta, a distance of 2,160 km (1,350 miles) as the peregrine flies. In the Americas Audubon, in 1803, was banding birds in a scientific manner, but we have to wait until 1890 when the Dane Mortensen fastened rings of zinc to the legs of starlings *Sturnus vulgaris*, each ring carrying a unique number, before individual birds could be recognised. In 1902 Bartsh banded black-crowned night herons *Nycticorax nycticorax hoactli*, also giving them a serial number. The usefulness of banding was soon appreciated and has made great strides since those early days. It was soon found that zinc rings did not last very long and the number and address carried on them quickly eroded in rainwater made acidic by dissolved carbon dioxide. Experiments have therefore gone on, and are still going on, to produce bands which the birds cannot get off, which do not inconvenience them in any way, can withstand rubbing on rock or shrubbery and will not dissolve in fresh- or sea-water. The scaly legs and large feet of birds have meant that leg bands are by far the best way of marking birds, but other methods have been attempted including neck bands on wildfowl and wing tags, also used on the Anatidae as well as on the flippers of penguins.

## Methods of Trapping

It is much easier to place bands on the legs of birds whilst still in the nest, but with many precocial species this is not easy and so it is a question of 'first catch your bird'. Methods used to this end have included drugging, herding, cannon netting, decoy trapping, mist netting, the use of heligoland traps and clap nets. A combination of these methods will enable a ringer to capture any species under investigation, and providing skill and sympathy are both present in abundance little damage is done to the birds.

DRUGGING

This is an unusual method and is illegal in the United Kingdom. It is done either by providing food containing a narcotic, or firing a drugged dart if the species concerned is large enough. The bird can then be caught, measured, weighed, sexed and banded prior to recovery and release.

### HERDING

This method is particularly suited to the ringing of wildfowl as the adults moult all their flight feathers at the same time and undergo a period of flightlessness before their young can fly. At this time they are all in family groups so they can be rounded up together. Birds carrying rings from previous years can also be caught at this time and so details of family life can be deduced.

### CANNON NETTING

This method also has been found useful in the capture of wildfowl, especially wigeon and geese which tend to feed on expanses of grass and also waders which gather in dense masses to roost or to feed upon estuarine mud flats. Nets are planted in areas favoured by the birds and they are shot into space by rockets which can be fired by remote control. This method has been used extensively in the United Kingdom by teams from the Wildfowl Trust, and experienced workers know how to set the nets and when to fire the rockets which carry the nets in a loop over the birds (*see* Fig. 97). It has to be accepted that this method does sometimes result in a few birds being killed but the knowledge gained from the results has meant that vital feeding grounds have been saved from land speculators and thus great benefits have been bestowed upon the species, though unfortunate for the odd individual. A new law restricting the use of this method to wildfowl, and laying down strict training procedures for operators, has recently been passed in the United Kingdom.

### DECOY TRAPPING

This is probably one of the oldest methods of all though now illegal in the United Kingdom except for duck trapping. In the days before battery hens and cheap fowls the occupation of bird catcher was a valued one in the community, as indeed is still the case today in parts of Europe where small birds, caught by liming the branches or in nets, are considered a delicacy. Decoys were used to capture waterbirds of all sorts, particularly ducks. A canal was dug, covered by an arch of vegetation and screened from view. A dog, usually a specially trained spaniel, was introduced at the opposite end to the watercourse and the birds would funnel down into the decoy (*see* Fig. 98). A door is closed at the end of the canal and the catch disposed of at will.

### MIST NETTING

An extensive length of fine meshed net is carefully sited by the ringing team

**Fig 97** Cannon netting. The inset shows the siting of the explosive charges which shoot the nets over the feeding or roosting birds

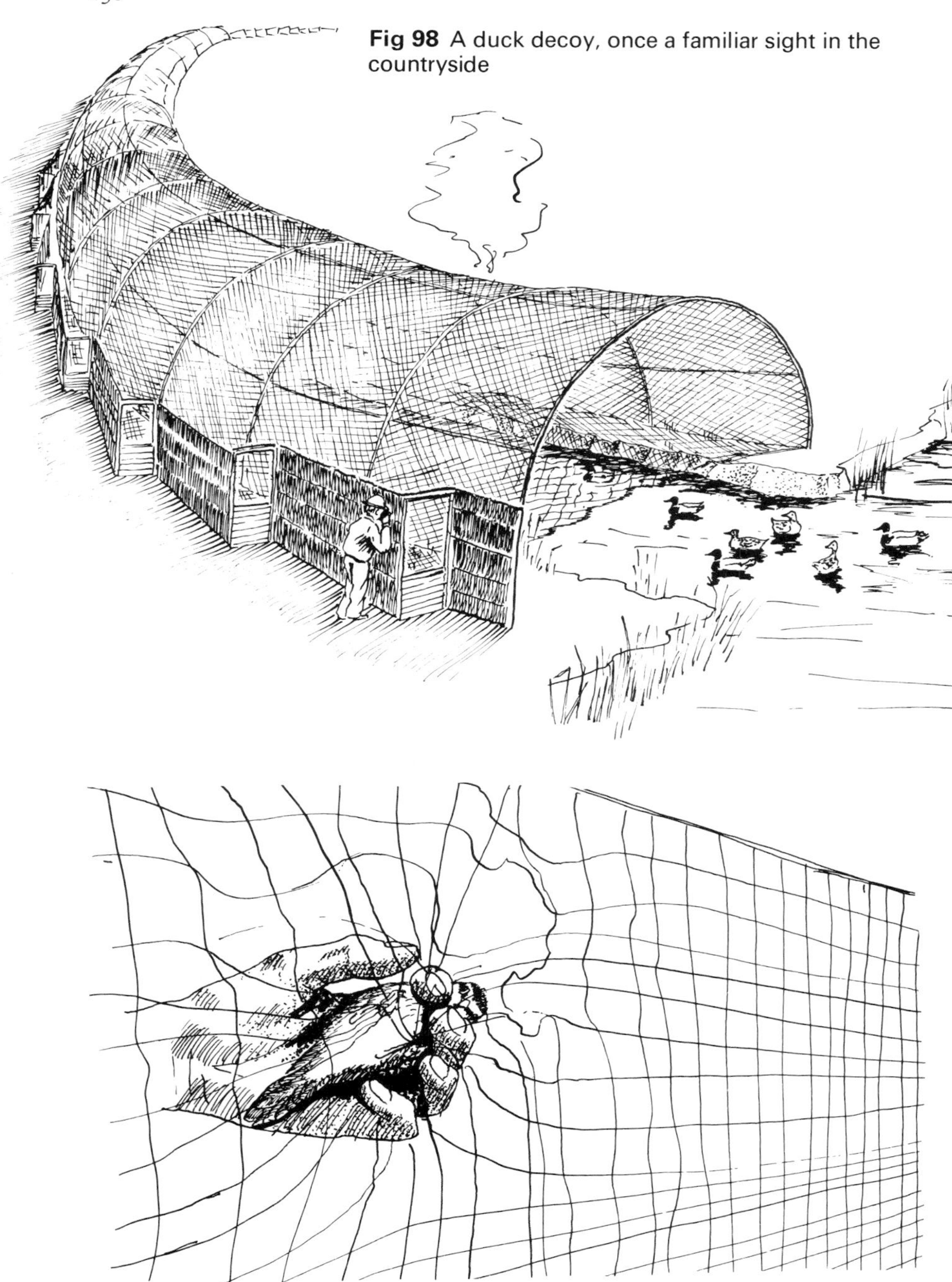

**Fig 98** A duck decoy, once a familiar sight in the countryside

**Fig 99** A small bird being removed from the mesh of a mist net

in an area where small migrants habitually gather (*see* Fig. 99). Large numbers can be caught in this way and casualties are extremely rare. The birds actually fly into the net which is fine enough not to be seen until the last moment when they thrust out their legs desperately clawing for grip only to hang there by their feet. Providing the ringing team work efficiently the birds can be dealt with very quickly and released without harm. Mist nets have the advantage of being usable at any time and in almost any situation.

HELIGOLAND TRAPS

These craftily designed structures get their name because they were first used by ornithologists working on the Atlantic island of Heligoland. The trap (*see* Fig. 100) begins as a wide funnel, the entrance being generously planted with shrubs favoured by birds both for cover and for food. The funnel gradually tapers inwards, being planted with suitable vegetation, and eventually the trap ends in a box. This has one side open, but which can be closed by a pulley operated from the entrance, often 20 or 30 m from the box. The ringing team allows its quarry to settle on the vegetation and then they begin a cautious approach. The birds scuttle down into the trap and then the ringers move rapidly and noisily driving the birds into the box. They are then 'processed' prior to release. This type of trap is very large and permanent and used only by bird observatories, but again results in surprisingly little damage to trapped birds and some resident individuals show so little concern that they are trapped two or three times each day!

CLAP NETS

This is a very simple device often used by ringers working on their own. A net is propped up by a stick with a string attached. Food is scattered under and around the net and when the bird moves beneath it the concealed observer pulls the string and the net falls on the bird.

Whatever care is taken in the trapping of the bird and whatever research has gone into the type of ring placed upon its leg, no results would be obtained if the rings were not recovered. Most countries now have their own central clearing house which receives the data from the ringer and collects the returns. The finder is then given the data originally provided by the ringer who in turn is given the details of where and when the ring was found and the possible cause of death. It is only by collating the data from thousands of returns that ornithologists can draw meaningful conclusions regarding bird movements. Some birds are either the victims of legal (or illegal) shooting or are large enough to be easily found and so a fair percentage of the rings placed upon such species are returned. With small shy birds, however, the bodies are not often discovered and so the percentage returns are very low. Some percentages are given in Table 11. It is not sufficient for one country's ringing scheme to operate in isolation since birds recognise no human barriers and it has been a regularly accepted thing for

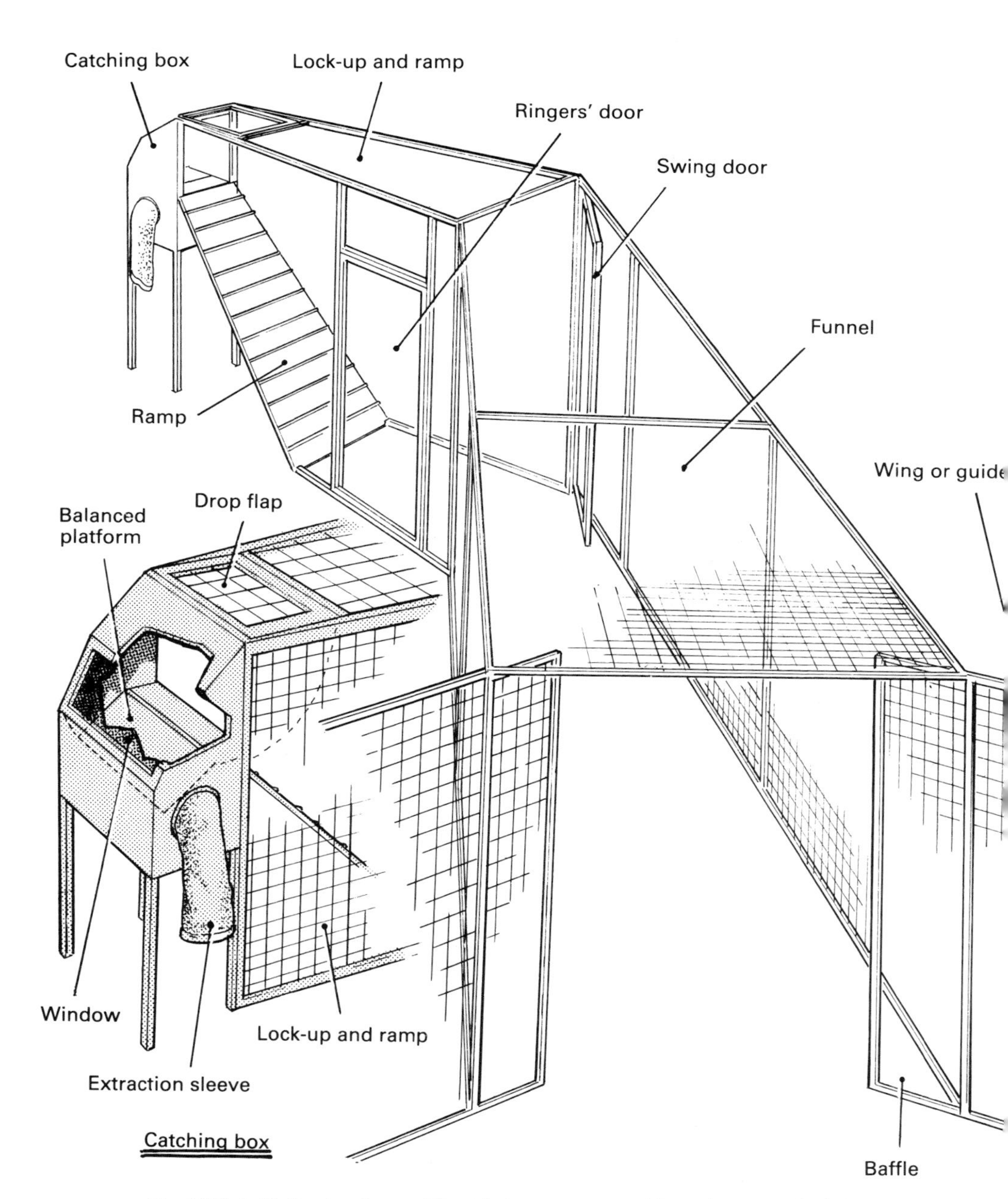


**Fig 100** A Heligoland trap. The birds are attracted into the tunnel of the trap by planting around the entrance with trees and food plants. The birds can be carefully removed via the extraction sleeve which is usually made of rubber

rings to be exchanged even between combatant nations. This is often done in the diplomatic bags taken to neutral countries where the exchange of scientific data proceeds smoothly. It is only as a result of extensive information that any deep understanding of migratory patterns will result.

**Fig 101** Heligoland traps are invariably used in coastal situations

Table 11
SHOWING THE PERCENTAGES OF BIRD RINGS RETURNED FOR SELECTED SPECIES
(Based on figures of the British Trust for Ornithology)

| Species | % Return |
|---|---|
| *Anser anser*–greylag goose | 24.3 |
| *Phalacrocorax carbo*–cormorant | 23.2 |
| *Anas penelope*–wigeon | 16.7 |
| *Anas crecca*–teal | 16.0 |
| *Accipiter nisus*–sparrowhawk | 15.0 |
| *Ardea cinerea*–heron | 14.7 |
| *Tyto alba*–barn owl | 13.0 |
| *Falco tinnunculus*–kestrel | 12.0 |
| *Fulica atra*–coot | 9.5 |
| *Scolopax rusticola*–woodcock | 7.6 |
| *Corvus frugilegus*–rook | 5.6 |
| *Sula bassana*–gannet | 3.8 |
| *Turdus merula*–blackbird | 3.2 |
| *Cuculus canorus*–cuckoo | 2.4 |
| *Parus caeruleus*–blue tit | 1.7 |
| *Motacilla alba*–pied wagtail | 1.6 |
| *Parus major*–great tit | 1.5 |
| *Fringilla coelebs*–chaffinch | 1.1 |
| *Hirundo rustica*–swallow | 0.7 |
| *Fratercula arctica*–puffin | 0.5 |
| *Sylvia communis*–whitethroat | 0.4 |
| *Riparia riparia*–sand martin | 0.2 |

## Migratory Patterns

There are many recognisable patterns to be considered including those determined by latitude, longitude, weather, altitude and the time of day.

### LATITUDINAL MIGRATION

This is the traditional north/south migration of temperate regions and typified by the European swallow *Hirundo rustica*; the work of Collingwood Ingram indicates that swallows home in on wintering feeding areas as well as on their breeding areas. British swallows, for example, make for southern Africa whilst birds from other parts of Europe winter in different areas of Africa. Typical of many north/south migrations is the presence of quite narrow channels called flyways through which streams of birds funnel in the spring and autumn of the year. In North America, for example, four major flyways are generally recognised, namely the Atlantic, Pacific, Central and Mississippian. These four routes are recognised by the Federal Fish and Wildlife Service and used in their banding programmes. Birds using them have no major geographical obstacles to overcome and so movements along them are very predictable as many birds find to their cost during the shooting season.

In Europe and Asia there are many more geographical problems to overcome, one of the toughest being the Mediterranean Sea which diverts many migrants, some passing down one side, others on the opposite side. Some funnelling does occur especially in the narrow area of the Bosphorus where soaring migrant birds of prey are very much to the fore during spring and especially autumn. Some flyways are very, very narrow, none more so than that used by the savannah sparrow *Passerculus sandwichensis princeps* which migrates from its breeding grounds on Sable Island in Nova Scotia in order to spend the winter in an area extending from Massachusetts to Georgia.

There will inevitably be some species which do not conform to the flyway pattern. This certainly applies to the Arctic tern *Sterna paradisaea* which breeds as far north as Greenland and northern Canada and winters in latitudes almost as far south, thus ensuring that it enjoys more daylight than any other species. Its autumn migration route may veer along the European coast and around Africa, before swinging off to Antarctica, whilst other individuals may make for South America then turning southwards again and thus reach Antarctica.

There are other equally fascinating species which in the course of time have extended into new areas, and the migrating birds then appear to follow this 'expansion path'. This has certainly been the case with the wheatear *Oenanthe oenanthe* which was originally restricted to Europe but reached Greenland and from there colonised Labrador. The migration route which is probably now genetically imprinted into the species takes it along this

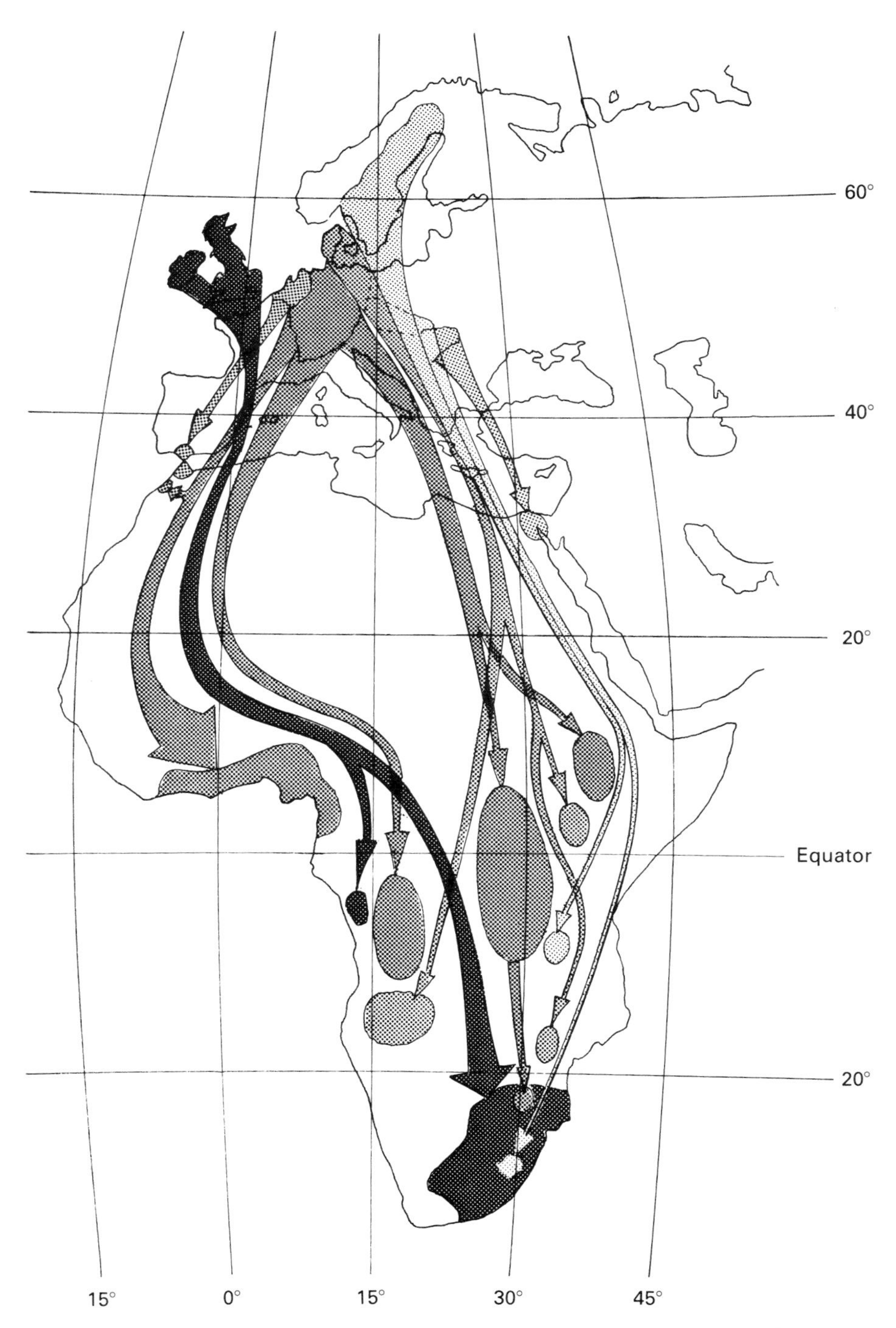


**Fig 102** The African wintering areas of swallows from different parts of Europe

ancestral route and those which follow it are more brightly-coloured and have longer wings than the normal wheatear and are deemed to constitute the sub-species *O. oenanthe leucorrhoa*.

### LONGITUDINAL MIGRATION

Not all avian migrants move north and south, there being several species migrating in an east/west direction. This has been demonstrated for the hawfinch *Coccothraustes coccothraustes* which breeds in Russia and winters in Japan and Korea. The British population seems not to find conditions too extreme and has settled down as a resident, although not very common, member of the avifauna. In North America longitudinal migrations have been observed in the evening grosbeak *Coccothraustes vespertina* which winters in New England and then moves west to breed in the state of Michigan. A similar pattern is found in the white-winged scoter *Melanitta fusca* breeding in central Canada and wintering either on the Pacific or Atlantic coasts, some birds moving due east, others due west, this being proved by an extensive programme of banding carried out by Cooke.

### THE EFFECT OF WEATHER ON MIGRATION

This is very difficult to measure with any degree of certainty since conditions can change with such rapidity. This did not deter David Lack who, in the early 1960s postulated that warm spring temperatures had a profound effect upon migratory movements. It has been found that when the European swallow migrates it moves into an area only once the temperature has risen to 10°C (48°F) and it seems that this particular isotherm also affects other birds such as the willow warbler *Phylloscopus trochilus*, but the flight of this species is more direct, and faster than that of the swallow because it does not depend upon a liberal supply of aerial insects for sustenance. Some workers have gone so far as to suggest a distinction between 'weather migrants' and 'instinct migrants' the latter being less affected by weather. In Sweden the swift *Apus apus* has been found to be particularly sensitive to temperature, and should this become unseasonally cold the species will temporarily move south, a phenomenon known as reverse migration.

Often high winds may develop during the course of a migration and this causes what Williamson sensibly referred to as migrational drift. Wind is a great hazard to birds and is one of the main factors responsible for the appearance of rare vagrants, and any birdwatcher worth his salt will keep a keen eye on the weather maps, looking for conditions that may produce rare vagrants. Pallas' warblers *Locustella certhiola*, for example, usually arrive in Britain from Asia on a high stretching across Europe and Asia, and American waders often turn up following low pressure in the Atlantic.

### ALTITUDINAL MIGRATION

Any land mass will have birds breeding close to the summits of all but the highest mountains, and as indicated in Chapter 5 they have the respiratory refinements capable of coping with quite rarified atmospheres. Areas suitable for summer breeding are often an impossible habitat during the winter, so many species have a migratory pattern which takes them up to the hills in summer and down towards the sea level in winter. Trends like this can even be detected in Britain where the mountains do not exceed 1,540 m (5,000 ft), both the curlew *Numenius arquata* and the lapwing *Vanellus vanellus* showing some traces of an altitudinal migratory pattern. In central California the story of the mountain quail *Oreortyx pictus* makes fascinating reading. Their nest can be situated at altitudes approaching 3,080 m (10,000 ft) where winter conditions become far too tough and in the fall small parties of birds set off on foot and migrate in single file dropping 1,500 m (5,000 ft) or so, only to return the following spring, again on foot, to the breeding grounds.

They say that there is an exception to every rule and it would seem that the winter movement of the blue grouse *Dendrogapus obscurus* up into the fir forests of the rocky mountains is totally illogical, but the cones of the fir are highly nutritious and the crop is reliable, and this coupled with the absence of predators makes this behaviour beneficial to the survival of the species. There is also some evidence to suggest that the dipper *Cinclus cinclus* may move upstream of frozen watercourses in order to reach areas of faster moving water which has not frozen.

#### DIURNAL AND NOCTURNAL MIGRATIONS

As indicated earlier in this chapter radar scans have shown that some species habitually migrate at night, but as early as 1886 Brewster was able to group birds according to their times of flight. His observations suggested that small passerines and larger but more secretive species tended to migrate at night. In Europe thrushes, warblers, cuckoos and woodpeckers migrate at night whilst birds of prey, crows and swallows migrate by day. With wildfowl, auks and waders it seems to make little difference, and movements occur both by day and by night.

## Orientation

Before speculations could be made regarding how birds navigate, ornithologists first had to build up a substantial pool of knowledge showing that birds could arrive at a given destination more often than was possible by pure chance—in short what was needed was the demonstration of a consistent ability to 'home'. It is difficult to carry out controlled experiments during a true migration and so most experiments have concerned themselves with the removal of incubating birds from their nests, transporting them some distance, releasing them and timing their return. These results are then assumed to apply equally well when birds are merely migrating.

Matthews working on gulls showed (1952) that in Britain the migratory lesser black-backed gull *Larus fuscus* had a much better 'homing' sense than its near relative the resident herring gull *Larus argentatus*. Ruppell in the 1930s transported over 300 starlings over distances varying from 48–656 km (30–410ml) from Berlin and found that over 30 per cent successfully 'homed' onto their nests, efficiency obviously decreasing with distance. Hinde in 1952 carried out tests on the homing abilities of the blue tit *Parus caeruleus* and the great tit *Parus major*, and found that they could not find their way home over distances in excess of 10 km (6 ml), but at the other end of the scale the Manx shearwater *Puffinus puffinus* is capable of phenomenal feats of navigation. A bird taken from its nest on the Welsh island of Skokholm was transported by aircraft to Boston, Massachusetts and released. It covered the 4,800 km (3,000 ml) in 12½ days in an east to west direction contrary to its normal north/south migratory route!

Midway Island's (Hawaii) population of Laysan albatrosses *Diomedea immutabilis* has in recent years been subject to a great deal of research because their flight paths are often a hazard to aircraft. In one classic experiment a bird released at Widby Island, Washington D.C. returned to Midway in just over 10 days at an average velocity of 504 km (317 ml) per day.

Thus we may safely conclude that birds *can* navigate very accurately, and we must now turn to the more difficult question of *how* they do it. It must be admitted that we have more theories than facts, including the suggestion that birds may be able to detect either the earth's magnetic field by means of their own cerebral compass, or the forces created as the world spins upon its axis (cariolis force). Most of the theories, however, involve the detection of geographical or astronomical landmarks, both dependent upon a highly developed visual apparatus (*see* Chapter 7).

Kramer proved conclusively that starlings *Sturnus vulgaris* could navigate accurately using the sun, and Matthews, who, like Kramer, published his work in the early 1950s, showed that Manx shearwaters *Puffinus puffinus* also navigated by the sun, and he further suggested that the birds were able to estimate the path followed by the sun across the sky (called the sun arc) and were able also to use this information to plot their course. This theory makes the assumptions that the bird can remember the sun arc angle at its 'home' territory, that its eyes can measure small angles, and that it also has its own built-in method of accurately measuring time, usually referred to as a biological clock. The work of both Kramer and Matthews stimulated other workers to copy their methods and apply them to different species, and also to adapt them in an attempt to come up with some plausible explanation to account for nocturnal orientation.

At night it seems most likely that birds follow the same clues as mariners, namely the stars. They do not seem to use the moon to any great extent and from experiments carried out on warblers and chickadees (titmice) by Sauer it seems there is an inborn as opposed to learnt ability to navigate by the stars. Birds reared in isolation from their ninth day were found to migrate in the correct direction following their release. Many species of bird have been placed in a planetarium, and subjected to either an autumn or spring star map. When the pattern shown corresponded to the correct season the birds orientated correctly, but they became confused if an autumn sky was presented in spring or a spring map in the fall of the calendar year. This shows that 'something' is built into the birds' system enabling them to recognise the correct star pattern. It was also shown that neither the planets nor the moon were used and this strongly suggests that the fixed stars were being used as points of reference. The birds therefore use either the sun or the fixed stars to establish the main direction of migration, whilst other landmarks and factors give the bird its final fix on its precise destination. Radar studies have shown that many birds, especially 'long range' seabirds, seem to be able to navigate when visual clues are at a minimum and so some other method must be operative, and the earth's magnetism, infra-red energy rising from warm sea currents, and even prevailing wind pressures have been suggested as possible clues. All this complex navigation, whatever its precise mechanism might be, must depend upon the bird having some accurate sense of time, a phenomenon which has been well-documented in many organisms.

## The Bird Clock

Put yourself in the position of a human navigator who is required to plan a long journey. Information required includes some sort of almanac of space, an accurate record of time, and an instrument to measure the angle of sun and/or the stars. We know how we do it, and we know how complicated it can be, and yet this sort of information is built into the make-up of birds perhaps even at cellular level. For example, starlings *Sturnus vulgaris* were trained to move in a certain direction and were rewarded by being fed. They seemed to work out their movements relative to the sun's position at daybreak. If captive birds were stimulated by artificial light switched on four hours before dawn the birds were four hours out all day. What the bird 'thinks' matters not—the experiment certainly proves that birds can tell time. Anyone who has ever put food out on a bird table at a fixed time knows that the guests are invariably punctual.

## Proximate Factors

We have so far considered why birds migrate and looked at the theories concerned with how they navigate—all that now remains is to consider what factors are involved in triggering the migration. These are technically termed proximate factors to distinguish them from ultimate factors such as optimum daylight hours for bringing up the young and plentiful food supply during breeding periods which rather account for why migration developed, than initiate movement in the individual bird. The two important proximate factors are the photoperiod and the build up of body fats. Great care must always be taken not to apply any conclusions reached as a result of studying one particular species to migrant birds in general. There are not only differences between species but also between individuals; bearing these limitations in mind we can now consider the two factors in question.

### PHOTOPERIOD

Rowan in the mid 1940s working in Edmonton, Canada was the first worker systematically to investigate the significance of daylight length by using electric lights artificially to increase the day length. He worked mainly on two species, a crow *Corvus brachyrhynchos* and the dark-eyed junco *Junco hyemalis*, and formed the opinion that the longer the day length the more exercise his birds got, and this raised the basal metabolic rate so that the sexual organs develop more quickly. It was, then, the hormones produced by the gonads which initiated the migratory flights. Other workers have shown that other glands such as the thyroid and pituitary are also important, and so all we can say at the moment is that the migration of some birds, at least, is initiated by an endocrine secretion or more likely by several secretions.

### THE BUILD UP OF BODY FAT

A long migratory journey demands vast reserves of fuel, and many species can build up fat deposits very quickly indeed. Some warblers whose normal weight is only about 10 gm ($\frac{1}{3}$ oz) can double this weight in a very short space of time and use it up just as quickly during the journey. The ability to cope with these great fluctuations in weight is a feature of many migrant birds and sedge warblers *Acrocephalus schoenobaenus* have been found to be so corpulent that the subcutaneous fat under the eyelids actually prevented the bird from closing its eyes. It has been suggested, but not proved, that it is this increase in body weight which acts as the trigger for migratory behaviour to begin.

*Chapter Ten*

# Behaviour

Ask a student of zoology to describe the behaviour of birds and you will often be submerged under a deluge of technical terms. These terms often cause the average birdwatcher to shy away from reading reports, papers and even books. Ask an amateur birdwatcher to define bird behaviour and you will usually be told that this is what the hobby is all about. The expert will often discard perfectly valid observations because of the use of 'amateur language' instead of the technical jargon. The purpose of this chapter is to explain some of the scientific terms so that enthusiastic amateurs will be able to translate the research papers and assist in improving the quality of their field notes and make them more valuable to the researcher.

Adaptive behaviour in living organisms including birds gives the species survival value, and this behaviour is influenced by several factors including the genetic make-up of the individual, its learning experiences, its physiological state (hormones, blood pressure, whether it is hungry or thirsty, etc.) its sensory input (what it sees, hears, tastes, feels, etc.) and what actions the bird is capable of performing from the muscular point of view.

## Historical observations

The history of our knowledge of bird behaviour may be said to have begun with Aristotle who studied breeding, locomotion and aggression, some of his field observations being very accurate, but interspersed with some really bizarre ideas such as 'when eagles grow old their beaks become more and more curved until finally they die because they cannot catch food with their curved beaks'. Such sweeping statements are, we hope, things of the past, but some workers are still inclined to note the initial and concluding portions of a behaviour pattern and then go on to invent the linking section. Even the great Darwin was at times inclined to be too anthropomorphic in his observations on behaviour and it was left to Lloyd-Morgan in the 1890s to approach the question in a purely objective manner.

By the 1930s E.L. Thorndike had suggested that when a stimulus was presented to an animal it responded to it, and should this reaction lead to a reward the behaviour would be retained but should it lead to punishment (or lack of success) then the behaviour pattern would tend to be discontinued. Thorndike and after him Skinner totally discounted the influence of heredity and explained all behaviour in terms of this stimulus/response theory. This held sway in North America, but in Europe Lorenz favoured an intellectual marriage between heredity on the one hand and learning from the environment on the other to account for behaviour, and coined the term ethology to describe this compromise. He worked extensively on jackdaws *Corvus monedula* and the greylag goose *Anser anser*. Tinbergen, a student of Lorenz, influenced his master by persuading him to accept some ideas from the States, and workers such as Hinde and Manning in Europe and Scott in America have also accepted the compromise, and modern work is aimed at understanding how and why a particular movement is initiated and how it is controlled.

## Fixed Action Patterns

Some of a bird's movements are obviously in direct response to the environment, such as the sight of a potential mate, rival or predator, whilst other movements seem to be totally independent of environmental influences. One of these seems to be the treading movement associated with nest building, and it often occurs whether suitable material is available or not. Lorenz called these fixed action patterns and they do not appear to be modified in any significant way by learning. One of these, the basic song pattern inherited by the European chaffinch *Fringilla coelebs*, has already been mentioned (*see* Chapter 8), and another concerns a piece of behaviour termed egg retrieving. A bird will retrieve an egg rolled from its nest in one particular way by pulling it backwards using the bill. This is efficient for a bird with a wide bill but very inefficient for those with a narrow bill, and yet all birds try this stereotyped method. Another example: there are about nine species of lovebirds (genus *Agapornis* of the parrot family), all closely related. Some species carry nesting material by stuffing it into their rump feathers. Others carry material in the bill. A hybrid between the two types behaves in a most peculiar manner. It can never get the material to the nest site because it retains both fixed action patterns, tucking the material into the rump feathers but refusing to let go with its bill! These patterns which are so vital in inter-specific communications seem on the basis of this and other evidence, to be inherited. The hybrid lovebird does, however, eventually succeed in building a nest but only after two or three years of frustrated effort. It makes preliminary movements towards the rump feathers and then carries the material in its bill. Thus fixed action patterns *can* be modified to some extent by learning.

## Imprinting

The work of Lorenz also suggests that a great deal of bird learning occurs during the first few hours of life by means of a process which he called imprinting, an idea he formulated by reference to earlier work by Spalding and Heinroth. He artificially incubated eggs of the greylag goose and on hatching the goslings followed their human keepers just as they would have dogged the footsteps of their natural parents, and as they mature they even

**Fig 103** Ducks will 'imprint' on the first thing they see after hatching. Usually it is their mother

direct their amorous attentions towards their keepers. Imprinting has been demonstrated in many species but this is hardly surprising since it is from their parents that young things get the protection they require in the natural world. The first thing a young bird sees on hatching will usually be a member of its own species—either a sibling or more likely a parent. Thus imprinting has a very obvious survival value and the rules of evolution will ensure its retention.

## Releasers

Similar species have similar behaviour patterns and these are often particularly obvious during courtship. The important question must now be posed—how do they differ? The answer often lies in quantitative behaviour—that is, how often the actions are repeated. If an action needs a level of stimulus before it is evoked what happens if two different species have different levels—one may perform a similar action more often, more strenuously or perhaps both. This may even have some effect on the evolution of new species (*see* Chapter 1). These exaggerated movements, often ritualised fixed action patterns, act as what Tinbergen has called releasers. In many ducks, for example, courtship preening results in a colourful area of the wing called the speculum being exposed, and this acts as a releaser initiating the next stage of the breeding cycle.

SIGN STIMULI

Some releasers are so exaggerated that they have been named sign stimuli. The red colour of a robin's breast for example is a sign stimulus which releases violent aggression in rival birds during the breeding season. At this time territory-conscious birds will react to any red object, as I found out during a spring day in Cumbria. Whilst helping to ring a family of jays in a wood I laid my anorak which had a red lining down on the ground and on my return found it being violently attacked by a robin which ignored my presence completely until I turned the coat over and the red lining disappeared. The bird then noticed I was present, gave its alarm call and fled.

SUPERNORMAL STIMULI

Supernormal stimuli have been noted in the meadow pipit *Anthus pratensis* and the herring gull *Larus argentatus*. Meadow pipits, like most birds, are attracted to an egg in the nest during the incubation phase of their breeding cycle. If the bird is given a choice between its own egg and a larger one it will choose the biggest; likewise if it is offered a larger clutch of eggs it will choose this in preference to its own clutch. It has already been mentioned (*see* Chapter 8) that the herring gull chick pecks at the red spot on its parent's lower mandible into order to persuade it to disgorge a meal. Experiments have shown that the supernormal stimulus in the form of a large red stick is preferred to the normal bill pattern. On paper these activities would seem to be detrimental to the species, but this is not the case. All that happens in the natural situation is that the signal selected is the most exaggerated one available, but there are no coloured sticks thrust at young birds by their parents and neither do birds in the wild state produce unlimited numbers of eggs, the clutch size being determined by the number present in the female's reproductive tract. It goes without saying that no bird can lay an egg of larger diameter than that determined by its oviduct and cloaca.

All that the average birdwatcher needs to understand is that all birds are bundles of fixed action patterns often modified a little by releasers, sign stimuli and by learning.

There are a number of aspects of bird behaviour, apart from the breeding cycle already covered in Chapter 8, which merit our attentions. These include sleep, group roosting, the mobbing of predators, the pecking order, bathing and preening. Finally we need to consider three aspects of bird behaviour which are causing particular interest, namely anting, the opening of milk bottles, and paper tearing.

## Sleep Postures

Sleeping positions do of course vary according to the species under consideration, but for creatures such as birds with their very high metabolic rate, the main problem during sleep is to minimise heat losses. The bill and

**Fig 104** The camouflage plumage of the treecreeper *Certhia familiaris* affords it a certain measure of protection even while roosting. A well used roost site, however, will have tell-tale marks of excreta

the legs are usually the only areas not insulated by feathers. Most adult birds turn the head backwards and tuck the bill into the scapular feathers, and many sleep perched on one leg tucking the other under the body, thus cutting down heat loss.

With regard to passerines many books indicate that during sleep specially adapted tendons lock the bird onto its perch and all it has to do is to lean at an angle thus tightening the grip rather like a noose. Some recent work now seems to discount this theory and observations of several species including tailor birds *Orthotomus* spp. have shown a frequent shifting of position. The flexing of toes and periodically tucking one leg into the body feathers shows sleep to be quite fitful, the bird's chief concern being always directed towards the conservation of heat. This does not rule out the validity of the 'auto-perch' theory, but it does mean there is a need to study the phenomenon in many other species.

## Group Roosting

This also seems to be directed towards conserving heat and has been observed in many numbers of species. Wrens *Troglodytes troglodytes* tend to make use of their nests all the year round, and during the British winter of 1978–79 I found 12 wrens frozen to death in a domed nest sited in a wall; there are records of at least 60 wrens using one roost. North American cactus wrens *Campylorhyncus brunneicapillus* also use old nests as roosts. Communal roosting has also been noted in short-eared owls *Asio flammeus*, rosy finches *Leucosticte arctoa*, and long-tailed tits *Aegithalos caudatus*, whilst during that infamous British winter of 1962–63 I found a mixed roost of blue and great tits *Parus caeruleus* and *P. major* in a nest box. Nine birds including six blue and three great tits were found to be dead, but whether they roosted together or whether the members of one species were already dead prior to the entry of the other is impossible to say.

## The Mobbing of Predators

Many birds have what amounts to an instinctive reaction to predators. The fact that a great variety of birds gather noisily to chivvy and chase owls has been known to bird catchers for centuries. Eagle owls *Bubo bubo* were used to attract birds into the range of hunters and trappers. Various corvids as well as small passerines were easily captured by placing a stuffed owl on a branch and then covering all the adjacent branches with lime. In America this method was successfully adapted: the great horned owl *Bubo virginianus* was placed near the nest of hen harriers, or as they are called in the States marsh hawks, *Circus cyaneus*, in order to catch them for banding. Mobbing behaviour has also been reported as directed against foxes, dogs, cats and grey squirrels, and does seem to be more common during the breeding season when eggs and young have to be protected. In 1961 Andrew found that the blackbird *Turdus merula* has a special mobbing call, used only when the attack has reached a frenzy. During this period even aggression between members of the same species is forgotten as they unite against the common enemy. Once this threat has been overcome the birds can return to normal and become territory-conscious again. Mobbing behaviour always entails some risk, but this must be worth taking if it serves either to alert the neighbourhood or as a 'refresher course' in predator identification for naive birds!

## Pecking Order

This behaviour seems typical of most animals, not only birds, and one dominant animal, typically a male, is able by sheer strength to have first choice in everything. This behaviour can easily be seen amongst domestic

**Fig 105** Mobbing behaviour: rooks *Corvus frugilegus* making life difficult for a buzzard *Buteo buteo*

fowl, but also occurs in flocks of wild birds, and I have seen it in operation in magpies *Pica pica* and lapwings *Vanellus vanellus*. Murton found that in woodpigeons *Columba palumbus* the establishment of a peck order was essential to their survival as a species. Anyone with bird tables can watch the pecking order in operation amongst the common birds which visit them. Once the breeding season commences even the poor old bird at the bottom of the pecking order is usually able to summon up enough aggressive energy to hold territory.

## Bathing

Bathing seems to be a regular feature in a bird's life and it does not restrict itself just to water, but often flaps about in dust, and really does seem to enjoy a good sunbathe. With regard to water bathing Simmons (1963) postulated five methods based upon his studies of passerines. These are by standing in shallow water, hopping in and out, dipping down from flight thus splashing water over their moving body, bathing in the rain, and finally by shuffling about amongst wet vegetation. Some species, including the hawk owl *Surnia ulala* have been observed bathing in snow, and I have observed similar behaviour in starlings *Sturnus vulgaris*, blue tits *Parus caeruleus*, and blackbirds *Turdus merula*. The inclination to bathe seems to develop quite naturally, soon after the art of flying has been mastered, and is always followed by careful preening and re-oiling of the feathers.

Dust-bathing has also been frequently observed, and in the house sparrow it has become a sort of communal ritual, a behaviour pattern also typical of the hoopoe *Upupa epops* and some game birds. Some of the dust-bathing areas seem well-known to the birds and they visit them regularly. The whole sequence of movements, both in water and dust is probably effective in reducing the irritation caused by external parasites such as fleas and lice.

Sunbathing seems to be enjoyed (or am I being too anthropomorphic) by all species, and it is thought that sunlight might possibly act upon the secretions of the oil gland producing vitamin D which could then be ingested during preening. Vitamin D is important for the correct formation of bone.

## Preening

This involves transferring oil from the preen gland to other areas of the body and has the obvious function of maintaining in good condition the feathers which are so essential to the bird's well-being. McKinney (1965) divided preening into three components which he called oiling, nibbling and washing, whilst earlier, in 1959, Goodwin noted that species such as the mallard *Anas platyrhynchos*, the purple martin *Progne subis* and the house martin *Delichon urbica* but especially the black tern *Chlidonias niger* were able to preen in flight—a sort of mobile repair shop.

A phenomenon known as allopreening, in which individuals groom each other, has been frequently noted and it has been suggested that this may well be a substitute for aggressive behaviour as a sort of displacement activity. It forms part of many courtship displays, which is what one would expect as the territory-conscious birds gradually have to accept the presence of the strange bird destined to be the breeding partner.

## Anting

This form of behaviour seems mainly, though not exclusively, restricted to the Passeriformes: within this order it is widespread and has been recorded in over 250 species. The green woodpecker *Picus viridis* of the order Pici-

**Fig 106** This roseate spoonbill *Platalea ajaja* has the advantage of a long neck when preening its feathers, but all birds manage to care for seemingly inaccessible spots

**Fig 107** The curious phenomenon of smoke anting or smoke bathing here demonstrated by a rook *Corvus frugilegus*

formes frequently ants, but the greatest exponents of the art are the corvids. The bird holds one or several ants in its bill and raises its wings, then rubs the ants along the feathers. This has been called active anting by McAtee (1938) to distinguish it from passive anting during which the bird just stands on an ant's nest and fluffs out its feathers, allowing the disturbed insects to reach the bird's skin. Dubinin, a Russian parasitologist has suggested that the process assists in the destruction of feather lice, and backed this up with some impressive statistics, whilst Whitaker working on a captive orchard oriole *Icterus spurius* thought that the burning feeling produced by the formic acid stimulated the birds in some way. I've no doubt that it did, but as it supports the activity as a useful function I tend to support the idea postulated by Dubinin.

An interesting variation in this behaviour, and which is still rather illogically called anting, concerns the use of other materials, particularly hot twigs from bonfires or other sources and glowing cigarette ends. These have been used by jays, *Garrulus glandarius*, carrion crows *Corvus corone* and rooks *Corvus frugilegus*, the hot smoking object being rubbed onto the feathers by the same ritualised method employed when anting. On occasions the smouldering objects have been carried by the birds into tinder dry forests and the eaves of houses, often with disastrous results. Finally starlings *Sturnus vulgaris* and house sparrows *Passer domesticus* as well as corvids have been observed 'bathing' in the smoke from chimneys, going through gyrations very similar to those used by anting birds. It is again presumed that this assists in the control of external parasites.

## Opening of Milk Bottles

This behaviour, so typical of the titmice was first recorded in 1921 when bottles were capped by cardboard tops with a perforated circle in the middle which could be pushed inwards to allow the entry of a straw. There is no doubt that the picking at this central point was learned and not instinctive, and the replacement of the cardboard cap with a bright shining coloured top has not stopped the spread of this behaviour, many birds even following the milkman to his delivery point.

Many species have been observed opening milk bottles including the robin *Erithacus rubecula*, song thrush *Turdus philomelos*, chaffinch *Fringilla coelebs*, great spotted woodpecker *Dendrocopos major*, and the starling *Sturnus vulgaris*. In the spring of 1980 I came across an injured carrion crow *Corvus corone* which, once cured, flatly refused to leave 'home' and now sits in a tree outside my study window. It periodically visits the doorstep and with a twisting movement of its bill rips off the cap and helps itself. It is even able to remove flat stones placed on top of the bottles to prevent its thieving. There is no doubt that the habit developed as a result of the inquisitive nature in some birds especially corvids and titmice, a trait which also probably accounts for the rather bizarre habit of paper tearing (*see below*). Doubtless, however, the food obtained as a result of the activities would reinforce the habit.

## Paper Tearing

There are a number of records, mainly from Britain, of titmice, mainly blue and great tits *Parus caeruleus* and *P. major*, entering houses and stripping wallpaper and damaging books. The phenomenon was especially prevalent in the 1957 'tit-irruption year' when population levels were enormously high and food supplies must have been under pressure. No completely acceptable theory has been brought forward to explain this, but it has been suggested that they could be eating the glue on the backing of the paper and bindings of the books. Some birds also chip away at the putty holding the glass panes in window frames, apparently for the linseed oil which it contains. Nothing is yet proved and I think that by far the most attractive theory is still to accept that inquisitiveness is very much a feature of the avian world just as it is in our own.

*Chapter Eleven*

# The Distribution of Birds

It seems most likely that birds arose in a tropical area corresponding to the region now occupied by Africa and India. From the Cretaceous period onwards (*see* Table 1, p. 14), waves of birds expanded from this centre rather like waves radiating from the area of a pond where a hurled stone strikes the surface. These waves are diverted by physical obstructions, a pattern also apparent in bird distribution. Several probable routes have been traced, including one into what is now South America (which has proved to be a dead end); others spread across Europe and Asia, and other areas.

In the early days of explorations it quickly became apparent to the intelligent and inquisitive expedition leaders that the wildlife differs from one area to another. Lists of fauna and flora compiled from Africa, South America and Australia differed greatly from those of Asia, India and Europe, and so the idea of zoogeographical zones did not come as a sudden flash of inspiration, but rather as a result of logical thought and observation over a long period of time. In 1876 Alfred Russel Wallace, the co-originator of the 'theory of evolution' with Charles Darwin, adapted Sclater's 1857 division of the world into zoogeographical regions, and applied the theory widely throughout the animal world, not just with reference to birds as Sclater had done. Although some modifications have been made since then, the scheme is basically the same as Wallace postulated. These regions are the Palaearctic, Nearctic (it has been the tendency in recent years for some workers to group the Palaearctic and Nearctic avifauna together under the title of Holarctic), Oriental, Ethiopian, Neotropical and Australian. Obviously there is bound to be some overlapping and confusion along the boundaries of the zones, and birds, being the most mobile of animals, are able on occasions to move from one zone to another, though examination of the birds of each region will reveal that many families have remained in situ. Recent work, particularly by D.W. Snow on the comparative avifauna of Africa and Europe, has revealed quite a degree of mixing, but geographical barriers may not be the only factors operating. A species invading a new zone, or being introduced deliberately (*see* Chapter 12), may not be able to 'muscle in' on a niche because it may already be occupied by a closely related but well-established species. Despite mixing, the avifauna of many zones is discrete enough to be obvious. Let us now examine each of the zones and discuss the typical birds associated with them.

## The Palaearctic

This region constitutes the northerly areas of the Old World. The boundaries are the Sahara and the Himalayas to the south, with sea on the other three sides. The whole of Europe is included and the zone embraces an area across Russia to the Pacific coast and to the west includes the Mediterranean coastal strip of Africa and a part of the north of Arabia. It is cut off from the Nearctic by a considerable stretch of sea except for the narrows of the Bering Strait. It is in land contact with both the Ethiopian and Oriental regions but is separated from the former by the Sahara and the latter by the Himalayas, both formidable barriers. Due to these land connections, however formidable, and the narrow Bering Strait, the Palaearctic has only one unique bird family, the Accentors (Prunellidae). The family consists of 12 species including the dunnock or hedge accentor *Prunella modularis* and the Alpine accentor *Prunella collaris*. The region is also very rich in members of the titmouse family (Paridae). Generally speaking, the birds found in the Palaearctic belong to very widely distributed families including birds of prey, wildfowl, finches, warblers and seabirds. Any consideration of bird distribution must naturally depend upon accurate identification, and nowhere has this been more apparent than in work carried out on wrens (Troglodytidae). Chapman and Griscom in 1924 listed 48 varieties (including species and sub-species) and plotted their distribution throughout Europe and Asia. By 1946 the work of other ornithologists including Mayr had shown that only *Troglodytes troglodytes* was a true wren, all the rest being timalids, otherwise known as babblers (Timaliidae), of which there are 282 species, all but one restricted to the Old World. The accepted theory at present is that one species of wren, *Troglodytes troglodytes*, spread into Europe via the Bering Strait, and this fits neatly into a pattern which accounts for the sub-species which have formed on St. Kilda and the Shetlands as discussed in Chapter 1.

On the whole, the climate over the Palaearctic region can be described as temperate, and habitats include coniferous and deciduous woodland as well as some tundra areas which are wet in summer and frozen in winter. There are also dry open steppe lands. Both these latter habitats are typified by extremely low winter temperatures.

### The Nearctic

This area includes the whole of North America and stretches as far south as the highlands of Mexico which form a barrier to the neotropical region, but they are what must be considered as a hazy boundary. This has not always been the case for at the beginning of the Tertiary period (some 50 to 70 million years ago) North and South America were physically separated, the land bridge in the area around what is now called Panama forming only about a million years ago. Following the formation of this bridge the exchange of fauna and flora would have accelerated but must have been somewhat limited by the differing climatic factors of the two zones. The

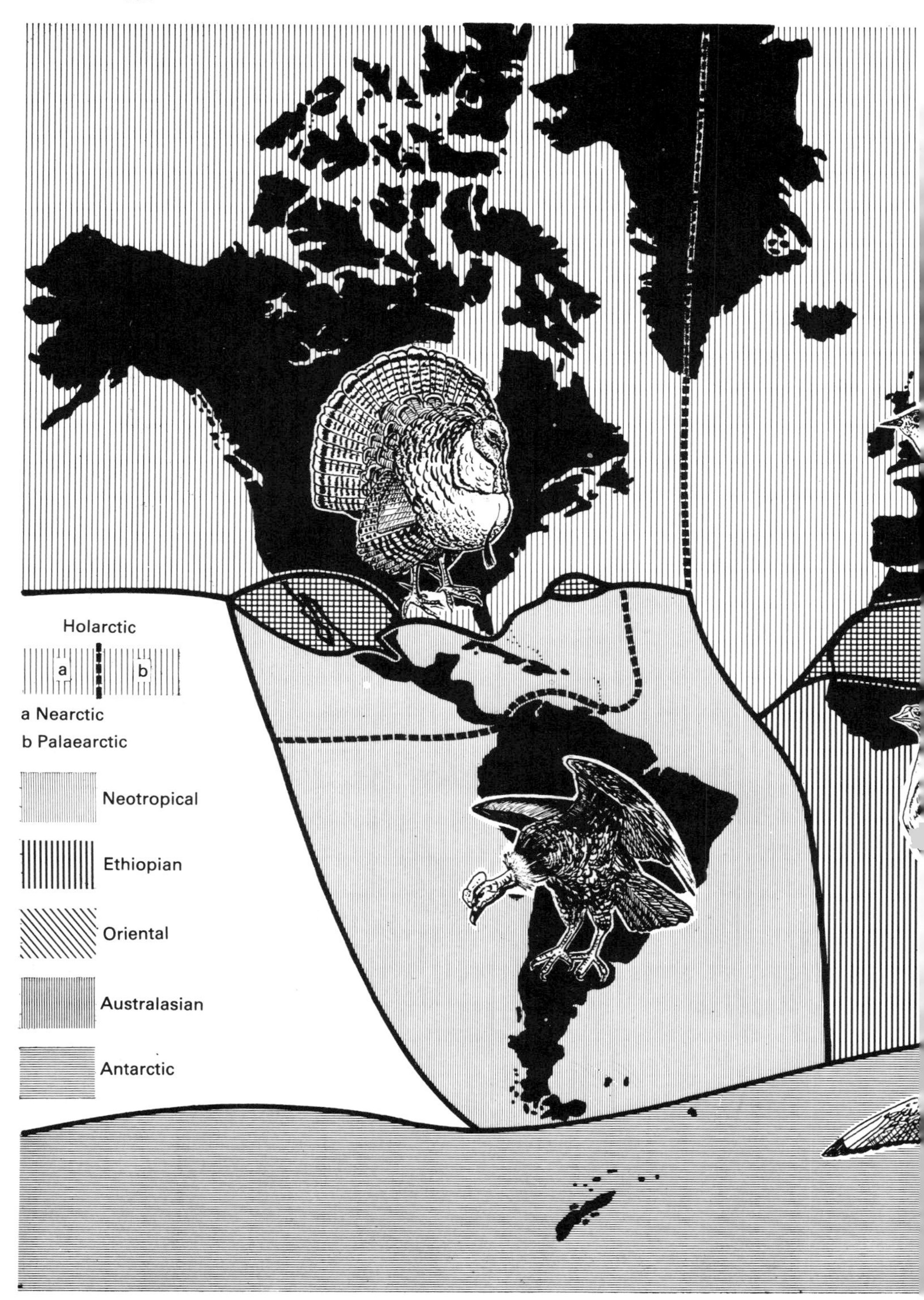
Holarctic
a
b
a Nearctic
b Palaearctic
Neotropical
Ethiopian
Oriental
Australasian
Antarctic

**Fig 108** The zoogeographical zones

Nearctic extends as far west as the Aleutian Islands and takes in Greenland to the East. We have already discussed the narrow channel of sea, the Bering Strait, which forms a link with the Palaearctic.

As far as the typical bird life is concerned there are far too many fringe areas to allow for the evolution of many unique species, the exception being the wild turkeys (Meleagrididae) which seldom stray into the Neotropical region and never into the Palaearctic. There are also quite a number of distinctive grouse as well as colourful cardinals and hummingbirds which have certainly spread from their original homes in Neotropical regions. Wood warblers are also abundant in the more temperate regions and wildfowl and waders around the more northerly regions. Mayr was of the opinion that mockingbirds (Mimidae), waxwings (Bombycillidae), New World vultures (Cathartidae), grouse (Tetraonidae) as well as turkeys and wood warblers originated in the Nearctic. It seems likely that the wrens originated in South America and reached the Palaearctic through the intermediary of the Nearctic.

The climatic conditions, and therefore by inference the vegetational zones, of the Nearctic region are very similar to those of the Palaearctic, this accounting for some workers suggesting one joint zone, the Holarctic.

## The Neotropical

This is a very neatly defined area, linked as we have seen with the Nearctic across the highland zone of Mexico, and most of this country is therefore included in the Neotropical zone, and from there it includes the whole of South America and the West Indies. It includes two of nature's gems, the towering Andes and the forested valley of the mighty Amazon. The birds are many and diverse, so much so that South America has been called by some the bird continent. It is perhaps the most vulnerable natural area in the world and, sadly, is being exploited at an unacceptable rate as man's greed for timber accelerates. Two orders, the rheas and the tinamous are totally confined to the region, and also typical are the crested toucans with their huge colourful bills, and the hoatzins. Deep in the forests of the Amazon are found the long-tailed macaws, and on the wooded Andean slopes are many busy hummingbirds, whilst soaring high on the thermals is the condor *Vultur gryphus*, also becoming much rarer. There are, however, a few gaps in the avifauna, only the quails representing the pheasant family, and songbirds are few in number, but this is more than compensated for by the great diversity offered by the presence of motmots (Momotidae), woodcreepers (Dendrocolaptidae), antbirds (Formicariidae), tanagers (Thraupidae) and ovenbirds (Furnariidae).

The region's climate is almost completely tropical and it is only at its southernmost limits that it begins to merge into a south temperate climate.

### The Ethiopian

The area embraced by this region includes Africa south of the Atlas Mountains and the Sahara, just taking in the southern corner of Arabia. In the north it contacts the Palaearctic but otherwise is isolated by sea. Some workers do not include the fascinating island of Madagascar in this region but I think that this is the best place to discuss it; here is a huge and ancient island situated 410 km (260 ml) off the coast of East Africa and which many believe was part of that continent during the Palaeocene or even the Eocene period (*see* Table 1, p. 14). Madagascar is very rich in birds belonging to three groups: some are endemic; others are African in origin; and just a few have penetrated from the Oriental region. There are four families which have been found nowhere else, and two of these are flightless. These are the mesoenatids which somewhat resemble rails, and the huge elephant birds which are now extinct. The remaining two families are forest-dwellers: the vangas feed mainly on insects and vertebrates such as chameleons, and the philepittas are fruit-eaters. There are certain gaps which are perhaps a little surprising: there are no woodpeckers or hornbills on Madagascar despite the fact that both families have wide distributions throughout both the Ethiopian and Oriental regions. The ostrich *Struthio camelus*, touracos (Musophagidae), and mousebirds (Coliidae) so typical of Africa have not reached Madagascar either.

So far as the mainland component of the Ethiopian region is concerned strong links can be found with the Oriental region, but the secretary bird, *Sagittarius serpentarius*, ostrich *Struthio camelus*, helmet shrikes (Vangidae), mousebirds (Coliidae) and crested touracos *Tauraco macrorhynchus* are all restricted to the region which is also rich in orioles (Oriolidae), birds of prey, cuckoos and woodpeckers. There are, however, very few parrots, pigeons or pheasants. Other interesting members of the Ethiopian avifauna include the desert-living long-winged sandgrouse (Pteroclidae) and the pratincoles and coursers (Glareolidae) whilst great diversity is seen in the Old World vultures (Accipitridae), shrikes (Laniidae) and larks (Alaudidae).

There are many similarities between the Ethiopian and Neotropical regions from the standpoint of physical geography; both have high mountains and grassy plains, huge river systems and evergreen forests. The Ethiopian, however, does not stretch so far south into the south temperate zone.

### The Oriental

This region includes the western islands of the Malay Archipelago, southern China, Indo-China and India. The limits are determined by the soaring Himalayas to the north, and on either side two mighty oceans, Indian and Pacific, form tangible limits. It is the south-eastern corner which spoils the neat arrangement, and a string of islands including Sumatra, Java, Borneo and the Philippines along with other smaller masses reach out towards the Australian region.

The Oriental region is justly renowned for its wide variety of trogons, crows, pigeons and cuckoos. Many species are brightly-coloured, and some show affinities with birds of the Ethiopian region especially the sunbills and hornbills. Just as in Africa there are few parrots.

It is the Phasianidae, however, the pheasants and fowls which should be synonymous with the Oriental zone, for here evolved the exquisite peacock *Pavo cristatus*, the argus pheasant *Argusianus argus*, and the jungle fowl *Gallus gallus* which is the ancestor of the domestic fowl. The region has only one exclusive family which consists of the fairy bluebirds (Irenidae). The climate is mainly tropical and has some similarities with the Ethiopian region but is nowhere nearly so rich either in the number of families or in the diversity they display.

## The Australian

This is an easily defined region as it has no land connections with any other region: it covers Australia, Tasmania, New Guinea and a few of the islands of the Malay Archipelago. New Zealand finds no place within this scheme, and as its 250,000 sq km (100,000 sq ml) lie some 1,600 km (1,000 ml) to the south-east of Australia we can hardly argue with this. I shall, however, discuss its avifauna as a rider at the end of this section because it does possess several interesting features.

The most striking family of birds found in the Australian region are the parrots (Psittacidae) which show much more diversity here than they do in the Neotropical zone. In Australia and New Guinea there are in the region of 85 species ranging in size from the 60 cm (2 ft) long cockatoos (*Cacatua* spp.) to the tiny pygmy parrots (*Micropsitta* spp.) which are just over 8 cm (3 in) long. Altogether 11 families are unique to the area, namely cassowaries (Casuariidae), emus (Dromaiidae), bowerbirds (Ptilonorhynchidae), birds of paradise (Paradisaeidae), kagus (Rhynochetidae), owlet-frogmouths (Aegothelidae), scrubbirds (Atrichornithidae), bellmagpies (Craticidae), mudnest-builders (Grallinidae), Australian nuthatches (Neosittidae), and the Australian treecreepers (Climacteridae).

With regard to climate a great deal of variation occurs. The north of Australia and New Guinea is tropical and a wide area is covered by rain forest. This contrasts sharply with the hot dry climate found in central Australia, and the further south you go the more temperate the climate becomes. There is, however, a marked divisional line separating the Australian from the Oriental fauna. It extends from north to south between Borneo, the Celebes (Sulawesi), Bali and east of Java to Lombok. This is aptly named Wallace's Line. In this case it is the barrier imposed by the sea rather than climatic factors which has affected faunal exchange. It also seems that birds which evolve in large land masses with more species to compete against are able to invade the smaller regions much more easily than can be achieved by those trying to move in the opposite direction. Thus more Oriental forms have reached Australia than the reverse.

## The New Zealand

Birds thriving in this temperate climate include three families unique to New Zealand. These are the kiwis (Apterygidae), the New Zealand wrens (Xenicidae) and the wattlebirds (Callaeidae). More flightless birds appear to have evolved here than in most other areas. Apart from the kiwis and a number of flightless rails there is also the kakapo or owl parrot *Stringops habroptilus* which now seems in some danger of extinction due to over-hunting. All other flightless birds, including some 20 species of moa, a flightless goose and a flightless wren are already extinct.

One species, a truly remarkable parrot called a kea *Nestor notabilis*, has quickly adjusted its feeding behaviour in response to the arrival of the sheep brought by the European settlers. The bird used to employ its huge and powerful bill to hack invertebrates out of their nooks and crannies, but now the kea tears huge holes in the backs of sheep and gorges itself on flesh and upon the fat which surrounds the kidneys.

New Zealand could have originated in one of two ways, either by gradually drifting away from Australia or by the erosion of a hypothetical land bridge during a rise in sea level. Either theory would account for its varied avifauna.

Whilst the faunas of some zoogeographical zones are very distinct there are sufficient 'blurred edges' to have caused many scientists, particularly Dunn (1922), to reject the theory altogether, his reason often being based upon fossil evidence showing that the distribution of many bird families has fluctuated widely over time. In the Miocene period (*see* Table 1 p. 14), for example, vultures and parrots occurred in southern France, and flamingoes were part of the Australian Pleistocene period—the list is long and impressive. This led to alternative methods of attempting to classify fauna, two of the most important being by reference to life zones and biomes.

## Life Zones

This theory put forward by Merriam in 1894 was something of a short-lived wonder and suggested that all plant and animal distributions could be explained by reference to average temperatures, often determined by altitude, over the area. By 1938 the work of Daubemire and others had somewhat discredited the theory but its spirit still ripples the sea of ornithological thought from time to time.

## Biomes

This method involves dividing up the world into areas occupied by the various dominant types of vegetation. Thus we have grassland, coniferous and deciduous woodland, tundra and desert biomes. This idea was pioneered in America by Pitelka (1944) and Kendeigh (1954). Once again compromise provides us with the best method of dealing with the problem of bird distribution since within each of the six zoogeographical zones we can recognise species typical of the different biomes: thus we can have Neotropical, Ethiopian and Australian deserts; Oriental and Neotropical rain forests; Palaearctic and Nearctic deciduous forests, and so on. This compromise often accounts for apparently discontinuous distributions of many bird families which are dependent partly on the zoogeographical zone but also upon their finding a suitable biome within the zone.

The modern view is that the concept of zoogeographical zones is a valid framework on which to build a picture of animal, and in our case avian, distribution. There is much work to be done on this subject, however, one of the most intriguing aspects of modern ornithology, and the work of Dr D. Snow in this field is always interesting. It is such a fascinating area because it is always changing, the climates of the regions being not invariable, and man's effect on the habitats around the world being ever more significant.

*Chapter Twelve*

# Birds and Man

This chapter can be viewed as consisting of three separate but closely interlinked topics: how do we harm birds, how do birds harm us, and what room is there for compromise? The first question can be considered under five headings which are destruction of habitat, exploitation of birds for food and amusement, deliberate transportation of species from one region of the world to another, pollution by oil and perhaps more importantly by other chemicals. On the other side of the coin those concerned with farming and forestry know birds as economic threats to their livelihood; they are also a hazard to those engaged in the dangerous business of aviation; and at times birds may even carry disease either to the human population itself or to our livestock and crops.

The answer to these problems is not to declare total war on all birdlife, but to make a careful study of their populations and, if need be, maintain these at acceptable levels. We have a duty as well, however, to monitor the populations of species in danger of extinction and do all we can to redress the balance. Both these aspects come under my definition of what conservation is all about.

## Man's Effect on Birds

### DESTRUCTION OF HABITAT

There are now very few regions of the world which remain unaffected by the hand of man, and they are becoming fewer and fewer as each year passes. Nowhere are these changes more apparent than in the so-called civilised countries. It is quite wrong to assume that all these changes are to the detriment of all birdlife. The felling of a deciduous forest, for example, must reduce the numbers of woodland-dwellers such as the corvids and the titmice, but the fields produced must allow an increase in the population of skylark *Alauda pratensis* and lapwings *Vanellus vanellus*. Man may decide to alter the environment once more by planting a monoculture of conifers. In the early stages the young plantations form ideal breeding habitats for birds such as short-eared owls *Asio flammeus* and hen harriers *Circus cyaneus*. In time, however, the mature forest offers succour to very few species because there is so little variation in food choice and therefore only relatively specialised birds such as crested tits *Parus cristatus*, capercaillie *Tetrao urogallus* and crossbills *Loxia curvirostra* can really thrive along with those predators at the top of the food chain such as the sparrowhawk *Accipiter nisus* and the long-eared owl *Asio otus*. Goodwin (1978) has expressed the opinion that it was unlikely that rooks existed in western Europe prior to

**Fig 109** The avocets are birds which have suffered from the reduction in wetland habitats. This is an American avocet *Recurvirostra americana*

man the farmer clearing the forests and ploughing his fields in which to plant his grain and thus producing an ideal rook habitat.

Thus it is easy to see that whilst some avian species suffer at our hands others find many advantages and thrive. One group of birds which have suffered greatly as a result of our activities are the marshland birds. Low lying swampy lands are ideal breeding areas for disease-carrying insects including mosquitoes which are vectors for malaria. The solution is to drain these areas and get rid of the pests whilst creating rich new farmland at the same time. The answer was perfect for all except the creatures living in these areas. Britain, for example, lost many breeding birds at this time including Savi's warbler *Locustella luscinioides*, ruff *Philomachus pugnax*, spoonbill *Platalea leucorodia*, bittern *Botaurus stellaris*, black-tailed godwit *Limosa limosa* and the avocet *Recurvirostra avosetta* now the proud symbol of the Royal Society for the Protection of Birds. Some of these birds have returned to breed in small numbers, and according to figures published by the magazine 'British Birds' and relating to the year 1977, 37-70 pairs of black-tailed godwits, about 12 pairs of Savi's warbler, and almost 150 pairs of avocets bred. The bittern also returned and breeds in East Anglia and Leighton Moss near Morecambe Bay but its population is once more giving cause for concern.

The effects of swamp drainage on birds has been noticeable in other parts of Europe and the range of such species as the white pelicans *Pelecanus onocrotalus*, cattle egrets *Bubulcus ibis* and the populations of other herons and ibises have shrunk alarmingly. Unless a halt is called soon many species will disappear altogether.

One of the advantages of our modern consumer society is the freedom with which exchange of goods can take place between one part of the world and another, but for trade to function efficiently docks must develop, and these are sited on estuaries, the very places where large numbers of birds have traditionally gathered. The birds are therefore pushed into smaller and smaller areas making them even more vulnerable to the shot, the shortage of suitable food and rising pollution levels. Modern industry needs power, and the vital electricity supplies are usually carried by wires stretched overhead between pylons which are often sited along migration routes. Large numbers of birds, especially swans, are killed by flying into the wires which are almost impossible to see during flight especially at night. Sympathetic officials have reduced the carnage by slotting large corks over the wires thus making them easier to see, but many birds still perish.

### FOOD AND AMUSEMENT

*Homo sapiens* has hunted birds for food since his earliest origins, but usually with so little thought for conservation that over-kill became inevitable as the human population increased, though many communities whose economy was based upon seabirds 'farmed' their 'stock' showing great concern for the future. The St. Kildans did it and the Faroese still do it by limiting catches to a given number, season or age class. One answer to the problem is to maintain an edible species in semi or complete domestication. The domestic fowl has been produced as a result of centuries of in-breeding of the red jungle fowl *Gallus gallus* naturally distributed through northern India, Malaya, China, Java and Sumatra. There are actually three other species of jungle fowl, the Ceylon *Gallus lafayettii*, the green *Gallus varius*, and the grey *Gallus sonneratii*, but our domestic chicken has been produced solely from *Gallus gallus*, which is naturally somewhat intermediate in size between the two extremes produced from it, the egg-laying machine of farmyard and 'battery farm' and the bantam. In the wild the males can be really fierce in defence of territory and this attitude has been fostered by centuries to produce the champions of cock fighting, a 'sport' which is unfortunately still far from dead.

The greylag *Anser anser* produced the farmyard goose and the domestic duck derived from the mallard *Anas platyrhynchos* providing solutions to man's problems in obtaining sufficient protein, but this does nothing to satisfy man's inborn urge to hunt. The American passenger pigeon *Columba livia* almost certainly became extinct because of man's primitive urge to kill, a fate which also befell the dodo and the great auk.

Much of the vegetation of northern Britain consists of heather moorland which is not a natural situation, but artificially maintained in order to support a high population of red grouse *Lagopus lagopus*. In other parts of Britain habitats are manipulated in favour of pheasants *Phasianus colchicus* or partridges *Perdix perdix*. Wholesale bird killing for food is still very much a feature of continental Europe whilst in Britain and America the shooting of wildfowl is still big business, and because of this great efforts are now made to conserve the quarry species carefully by laying down extensive close seasons and in some cases limiting the bags. Many wildfowlers now know how to recognise the rare species and those with less attractive flavours and leave them alone. Many birds, however, still fall victim to the unscrupulous.

Birds, to primitive people, were valued not only as food but also for making clothes which were perfectly insulated by the feathers to keep out the winter cold. Perhaps the best example of this today is the 'farming' of eider ducks which is very much a part of Iceland's economy. The eider *Somateria mollissima* lines its nest with a thick down plucked from its breast. During, and again at the end of, the breeding season the down is harvested and used to line sleeping bags and bedding thus giving us the 'eiderdown'. In some parts these coverings have gone out of fashion, being replaced by the duvet, which is the French name for the eider. At one time, not so long ago, the feather industry almost reduced a very large number of

**Fig 110** An eider duck *Somateria mollissima* nest, showing the lining of down

bird species to total extinction. It seems that the practice of ladies wearing gaudy feathers may have been brought back to Europe at the time of the crusades, and the history of this practice has been traced by Doughty (1975). The New World settlers, on the other hand, seem to have frowned on the practice and a minister named William Morrel in 1620 had plenty to say about this and he set down his views in verse.

The fowles that in those bays and harvests feede
Though in their seasons they doe elsewhere breed
Are swans and geese, herne, pheasants, duck and crane
Culvers and divers all along the maine:
The turtle, eagle, partridge and the quaile,
Knot, plover, pigeons, which doe never faile
Till sommers heat commands them to retire
And winters cold begets that old desire
With these sweete dainties man is sweetly fed
With these rich feathers ladies plume their head
Here flesh and feathers both for use and ease
To feed, adorne and rest there, if thou please.

Initially Europe had no religious doctrines to restrict extravagance and the wearing of feathers seems to have reached its peak in the court of Marie Antoinette and Louise XVI of France which came to a grizzly end with the revolution of 1789. The French example stimulated other European courts and the demand for ostrich and peacock feathers soared as did, of course, their price. In moved the speculators of the millinery trade, and the search for birds with attractive feathers was on. By the 1860s some conservation-minded folk in both America and Britain began to complain about the feather trade, and it is interesting to note that both the Audubon Society in America and the Society for the Protection of Birds (now the R.S.P.B.) both had their origins in the anti-feather movement. Eventually they won their case and the tradition of wearing feathers dating back to the crusades and before had vanished.

Birds and man retain another link, however, with the saracens of long ago. There are conflicting reports regarding the origins of falconry, but it certainly did not originate as a sport. In North Korea there are still falconers who make a living selling birds killed by their hawks; the food is traditionally shared (however unequally) with the birds just to keep them sharp. Britain has a continual tradition of falconry up to the present day, whereas in most of the countries of Europe the practice lapsed for a time, before being revived. In 1927 the British Falconer's Club was formed and is still thriving: most, but unfortunately not all, of the birds used in falconry are bred in captivity. The tremendously high price paid for an efficient bird, particularly a peregrine, has brought many unscrupulous people into a conflict with British conservationists and the law of the land. It is true to say that by far the majority of falconers have a great regard for wild birds and do as

**Fig 111** Examples of some of the bizarre trappings of falconry

much to conserve wild species as any other group of sportsmen, and it is a few greedy individuals hell-bent on profit who do the damage both to bird populations and the image of the sport.

There are some areas, particularly Ethiopia, India and Pakistan where birds are not hunted, but left alone, perhaps because of religious taboos. However, sheer levels of human population can have a more subtle, yet equally destructive, effect on some species.

INTRODUCTIONS

It was the practice of many early explorers to bring back examples of the more exotic flora and fauna which they observed on their travels. Many of these escaped or were deliberately released and this has led to the mistaken impression that all you have to do is to collect a cage full of your chosen species, dump it in another zoogeographical zone and hey-presto, out they jump and populate the new habitat. By far the great majority of introductions fail, since they could succeed only in a suitable climate and habitat.

It seems that the most successful introductions have been birds whose native populations were fairly high and so they were used to competition. Also they were used to living alongside mankind and could co-exist with the associated noise, buildings and fast-moving transportation systems. The starling *Sturnus vulgaris*, and the house sparrow *Passer domesticus* have both successfully established themselves in North America, Australia and New Zealand, but only after several separate groups had been transported from Europe. Other attempted introductions have been less successful; perhaps

**Fig 112** Canada geese *Branta canadensis*

the most surprising has been the rooks' *Corvus frugilegus* failure to find a niche in New Zealand. In Britain many attempts were made to introduce the American robin *Turdus migratorius*, but this failed obviously because its habits and habitat are almost identical with Britain's resident and resilient blackbird *Turdus merula*. There was thus no ecological niche suitable for it and *T. migratorius* was over-whelmed. Other more successful introductions have been the pheasant *Phasianus colchicus*, the red-legged partridge *Alectoris rufa*, the Canada goose *Branta canadensis*, the ruddy duck *Oxyura jamaicensis*, and the little owl *Athene noctua*.

The pheasant is often stated to be, along with the rabbit *Oryctolagus cuniculus*, one of the species introduced by the Romans. It now seems pretty certain that both these suggestions are erroneous, the rabbit being a Norman immigrant, the pheasant coming in just a little before the Norman conquest, probably from the Caucasus. Goodwin (1978) is of the opinion that the pheasant might just possibly have been a very rare native of Britain, its ranks being swelled by a series of introductions, often involving sub-species, from the eleventh century onwards. The red-legged partridge was introduced into Suffolk from south-west Europe round about 1780 to add variation to the sporting scene which was just beginning to gather momentum. The 'Frenchman' as it is called is now well-established in eastern England, and is occasionally encountered in other areas of Britain despite the fact that it is well to the north of its usual range.

Wildfowl have always found favour with aviculturists, mainly for their aesthetic value. The Canada goose *Branta canadensis* was known in Britain in the seventeenth century, but fairly large scale introductions commenced only in the eighteenth century as country houses with their parks and lakes mushroomed. Many of the feral flocks are still found round these country parks during the winter; in summer they tend to pair off and drift away to breed, only to return in the autumn, their fast maturing families still in tow. The species seems to have lost any migratory urge which it shows in its native North America. Amongst the ducks, the delightful little mandarin *Aix galericulata* from the Far East is also beginning to breed ferally in Britain, especially in Berkshire where it finds few competitors in its chosen habitat along fairly swift streams running through woodlands. It may well succeed in establishing itself to the same extent as the ruddy duck *Oxyura jamaicensis*, a North American species originally brought to the Wildfowl Trust's collections at Slimbridge, Gloucestershire. In constrast to many species it was found to be more proficient in raising its own young than with any of the artificial methods available at the time. Some of the young produced avoided having their wings clipped and escaped. From these escapes we now have perhaps 500 birds at large, many in Cheshire, and the population is rising rapidly.

The ring-necked parakeet *Psittacula krameri* is a more recent escape, and is already causing concern to fruit growers in south-eastern England.

The story of the little owl's *Athene noctua* introduction into and colonisation of Britain is perhaps the best documented of this type of experiment. In 1842, five birds were released in Malton Park, Yorkshire, but, as is often the case, the exercise failed totally. Much more successful, however, were the efforts of Meade-Waldo in the late 1870s and of Lord Lilford in Northampton in the late 1880s, and from then the little owl has never looked back. By the time of the British Trust for Ornithology's Atlas Survey of 1976 the breeding population was estimated at between 7,000 and 10,000 birds. It succeeded because it found a suitable niche in the ecology of Britain since none of the other owls—barn *Tyto alba*, tawny *Strix aluco*, long-eared *Asio otus* or short-eared *Asio flammeus*—had any preference for stone walls.

Most countries of the world have a long list of failed introductions, and a short, but often impressive, list of species which have successfully broken new ground. This is certainly the case in the United States, Australia and New Zealand.

In the States several attempts were made in the middle of the nineteenth century to introduce both the house sparrow *Passer domesticus* and the starling *Sturnus vulgaris* but they met with failure until 1890 when 60 starlings were introduced to Central Park, New York, and another 40 were released the following year. They have increased to pest proportions in some regions, and so indeed has the house sparrow.

*Passer domesticus* was also introduced into Australia and once more has proved a good colonist, but it has still not penetrated very far into Western

Australia. The blackbird *Turdus merula* was introduced to the Melbourne district in the 1860s as was the song thrush *Turdus philomelos*, but it is again the starling which has led the way and is abundant in many towns and has even managed to colonise Tasmania. New Zealand has also received its share of birds from the Old World, most of which she could well have done without; for better or worse she now has starling, house sparrow and the chaffinch *Fringilla coelebs* which is very successful here; the song thrush and blackbird are both more successful in New Zealand than they are in Australia, but this is predictable since the habitat is similar to that of Europe. Other introductions include greenfinch *Carduelis chloris*, goldfinch *Carduelis carduelis*, yellowhammer *Emberiza citrinella*, and the dunnock *Prunella modularis*, which, as we have seen, is native only to the Palaearctic zone. The redpoll *Acanthis flammea* has also proved a successful pioneer in New Zealand where it is causing concern to fruit growers.

Now that the damage which can be done to native fauna by introductions is appreciated we may hope that the practice may cease—regretfully I doubt this, but at least more care may now be taken to confine captive birds to their aviaries.

### OIL POLLUTION

In 1907 a schooner was wrecked off the Isles of Scilly and released thousands of gallons of oil. Since then the problem has got worse, and on a worldwide scale. Oil tankers are huge floating potential disaster areas, but there is an even greater threat as supplies become more difficult to extract and spills much more difficult to control from fields under the sea. Oil pollution from tankers, drilling areas and pipelines are accidental, but not all spills are of this type. Some are quite deliberate, being due to tanks being cleaned out at sea. The development of the huge tankers has meant that when spillages do occur they are much more serious, one of the most infamous being the *Torrey Canyon* in 1967 which killed thousands of birds including 8,000 guillemots *Uria aalge* and razorbill *Alca torda*, and also polluted the holiday beaches of Cornwall. Sometimes, however, it is not so important how much oil is spilled but where it is spilled. Swennen and Spaans (1970) reported upon the effects of the spillage of only 150 tons of residual fuel oil in the Dutch Waddenzee during February 1969, a time when many seabirds concentrate in the area; no fewer than 41,000 birds perished.

It should not go unnoticed that the huge oil installations off northern Britain are often close to large and important seabird concentrations, and significant numbers of great northern divers *Gavia immer* and eiders *Somateria mollissima* have already been killed. There have already been some accidents in this field of exploitation of the world's resources—in January 1969 a blow-out occurred from an undersea oil drill in the Santa Barbara channel off California. Each day for 12 days tons and tons of crude

**Fig 113** An all-too-familiar sight: an oiled guillemot *Uria aalge*

oil spilled to create a slick which stretched for 1,300 km (800 ml) into the Pacific and polluted over 30 beaches; all life suffered, and almost the total population of 4,000 western grebes *Aechmophorus occidentalis* was destroyed.

Attempts at legislation to combat the problem were first made in Britain in 1922 and this prohibited the discharge of crude oil into her territorial waters. This did not work then, and it does not work now, because there are too many vessels sailing under flags of convenience which are prepared to risk washing out their tanks at sea knowing how difficult the offence is to detect, especially if done under the cover of darkness. By the 1950s the number of seabird deaths was beginning to disturb people of influence, and the result of their protests was the setting up of an Advisory Committee on Oil Pollution of the Sea. In October 1953 this committee organised a conference which was attended by 28 countries. The result of their deliberations was the 'Oil in Navigable Waters Act' and this was amended in 1962 to prevent any discharge of oil into the Baltic and the North Sea. The oil companies played their part by the introduction of what has become known as the 'load-on-top' system. A sort of slop tank is built in to tanker design and a mixture of bilge and oil flushed out of the tanks can be pumped into this instead of into the sea. In time the mixture in the tank settles out with the oil on top and the more dense seawater below. The 'lot slops', as the floating oil is called, is retained whilst the slightly polluted seawater is discharged. At the loading port the lot slops can be mixed with the next lot of crude oil shipped aboard.

Although the existing laws are nowhere nearly tight enough with regard to oil dumping at sea, it has been estimated that twice as much oil reaches the sea from land based activities, as from shipping. When an oil spillage kills birds on an estuary conservationists glare out to sea and curse ships, when perhaps they should look upstream at the throbbing motorways and the heavy industries.

So much for oil spillages, but how precisely does oil affect the birds? Obviously the oil itself is toxic, and when coated on the feathers is preened off by the bill and some of this is swallowed, irritates the gut and finally kills. Oil also clogs the barbs of the feathers and prevents them from acting as a barrier against water and cold. This leads to waterlogging and the birds usually sink, only a few actually being washed ashore. At one time it was thought that washing the affected birds with detergent was the answer, but this is now known to be less than one per cent successful because the detergent itself robs the feathers of their waterproofing qualities, so that released birds quickly succumb. Even if a successful method of treatment was discovered how do you go about the task of treating 20,000 oiled seabirds? This is a case of prevention not only being easier than cure but also considerably cheaper. What is needed is a deep understanding of seabird populations and their annual cycle of movements, especially those colonies sited near oil installations. This coupled with stringent control of deliberate oil spillages on land and at sea should keep this emotive problem within acceptable limits, and will allow us both to obtain vital energy supplies and keep our birds.

## CHEMICAL POLLUTION

Following World War II the use of pesticides began to gain momentum, and by the beginning of the 1960s the organochlorines were being used as seed dressings. The two most toxic of these compounds are dieldrin and aldrin, and in Britain during 1960 and 1961 huge numbers of birds perished as a result of their use. A related substance, an organophosphorus called parathion was killing Dutch birds. D.D.T. is a member of the organochlorine group and a relatively mild one, but even its effects can be devastating. It was used at Clear Lake, California to reduce the population of midges which were annoying tourists. The effect of the treatment was not only to kill midges but also to reduce the 2,000 strong breeding population of western grebes *Aechmophorus occidentalis* to about 60 birds, and even those which attempted to breed failed. It became obvious that the harmful chemicals were becoming incorporated into the food chains. Some of the sprayed chemical entered the water to be picked up by invertebrates which were eaten by fish; these in turn formed part of the diet of the western grebes. Since this time it has often proved possible to use birds as indicators of pollution levels which not only affect the birds themselves but also us because we are also at the top of many food chains.

The work of Ratcliffe and of Newton on birds of prey in Britain, and Peakall in the United States has shown that organochlorides cause the liver to produce substances which adversely affect the functioning of oestrogen which in turn controls the calcium production associated with shell formation. This causes the birds to lay eggs with thin shells which break during incubation. Since the late 1960s eggshell thinning has been noted in the European sparrowhawk *Accipiter nisus*, American sparrowhawk *Falco sparverius*, peregrine *Falco peregrinus*, herring gull *Larus argentatus*, as well as the brown pelican *Pelecanus occidentalis* and the white pelican *Pelecanus onocrotalus*. The method used to estimate the extent of this thinning can be seen by reference to Table 12.

Table 12
EXTENT OF VARIATION IN SPARROWHAWK EGG-SHELLS FROM DIFFERENT REGIONS OF BRITAIN

| | **Until 1946** | | **From 1947** | | |
|---|---|---|---|---|---|
| | **Number of eggs examined** | **Mean shell index* (± standard error)** | **Number of eggs examined** | **Mean shell index* (± standard error)** | **Extent of change** |
| South eastern England | 68 | 1.42 ± 0.01 | 129 | 1.12 ± 0.01 | −21% |
| Other regions | 230 | 1.43 ± 0.01 | 150 | 1.23 ± 0.01 | −14% |
| All regions | 298 | 1.42 ± 0.01 | 279 | 1.18 ± 0.01 | −17% |

*Calculated by the formula $\frac{\text{Weight of shell (in milligrams)}}{\text{Shell length} \times \text{breadth (in millimetres)}}$

*From Ratcliffe, 1970*

These chemicals are now gradually being brought under control, but it is not only agricultural chemicals which can cause damage to the environment, and industry has also added to the problem usually in the form of chlorinated biphenyls (PCBs). These have been in extensive use since the 1930s and are contained in plastics, paints and oily lubricants. PCB-based pollutants have also been proved capable of entering food chains and have been found in the tissues of marine molluscs, fish and many species of birds found at sea including wildfowl, waders, gulls, auks, terns and petrels. Heavy metals such as zinc, arsenic, lead and mercury have also been detected in seabirds. These chemicals reach the sea via rivers or from coastal areas, and are extensively used in the production of both fungicides and paper. Once again the birds are useful indicator species and seabird deaths from such substances must mean the presence of a threat to human life.

Already high levels of mercury have caused deaths in Japan from what has been called Minimata Disease contracted by eating contaminated fish. What is even more disturbing is the more subtle effects of PCB poisoning in some seabirds such as the shag *Phalocrocorax aristotelis*. The poisonous residues tend to accumulate in the stores of body fat. These may not be drawn upon until times of food shortage when the chemicals are released into the blood stream and may well be lethal. If this can happen to an indicator species then why not us? If all we need is a selfish reason to protect our wildlife then this surely is it.

Another potential pollutant which needs careful monitoring is the release by accident or by the deliberate sinking at sea of radio-active waste. We need power, but we also need to be careful and the monitoring of fauna and flora in a search for unusual genetic effects must be carried out whatever the financial cost.

It is not just the sea which is becoming contaminated, and many streams, rivers and lakes are heavily polluted, and even the more innocent material such as rich fertilisers can cause problems, as can raw sewage discharged without adequate treatment. Normally a clear stretch of fresh water is defined as oligotrophic, i.e. lacking nutrients, but it becomes eutrophic when the nitrates and phosphates begin to build up. These accelerate plant growth, this in turn builds up the animal populations, and these soon begin to exhaust the oxygen levels. In hot weather (or if hot water is discharged by industry—termed thermal pollution), since oxygen is more soluble in cold water than hot the oxygen level may be insufficient to support life, and so the whole fabric of the ecosystem is destroyed and the birds, so much a vital part of it, are destroyed with it.

Atmospheric pollution is another factor to consider, and it has been suggested that this is felt more forcibly by human populations than by wildlife, the large numbers of people dying in the London smogs of the 1950s and the hazards of breathing the air of Los Angeles and Tokyo being especially notorious. But it is not accurate to say that the total environment is not affected by apparently localised pollution; sulphur dioxide from western Europe, for example, has been carried by prevailing winds to Sweden where it has caused great damage to the growth of the conifer forests and the specialised avifauna which depends upon them.

One way of proving that birds are affected is by conducting an experiment in reverse, by cleaning up the pollution and watching the wildlife return. Following the 'Clean Air Acts' in London the smogs became a thing of the past and house martins *Delichon urbica* returned in increasing numbers to breed in the city. A similar effect has been noted, returning to the subject of polluted waters, during the 1960s and 1970s when a great effort was made to clean up the Thames, a costly but dramatic improvement resulting. Down went the sewage input due to the opening of new treatment plants, up went the levels of oxygen, followed by dramatic population rises of both plants and small animals. This was followed by the return of fish and birds—thousands of healthy birds back on what was once one of the world's polluted rivers in less than 20 years. This improvement has been well-documented by Harrison and Grant, and by Wheeler. The lesson seems to be perfectly clear—monitor large, colourful indicator species such as birds, keep them free from lethal pollution and our reward will be twofold. We will have an abundance of lovely creatures to look at, and a healthier environment for ourselves.

## The Effect of Birds on Man

### BIRDS AS ECONOMIC PESTS

Man's agriculture by its very definition is unnatural, and if a single species is planted in an area which once supported many others, it is vulnerable to disease which will spread much more quickly through a monoculture than it would through a mixed community. Should we plant orchards full of large succulent fruit arising from sappy springtime buds there will be a bird species which will take advantage of it. Either the trees must be protected by expensive netting or by expensive deterrent chemicals, bangers, scarers and other devices. The possible ravages of the bullfinch *Pyrrhula pyrrhula* on fruit trees in Britain has long been realised. Damage to crops of apples, pears, cherries and gooseberries can ruin a farmer, and as long ago as Tudor times an Act of Parliament laid down that one penny was to be paid out of the parish accounts for each bullfinch killed. Other passerines found guilty of gobbling up seed crops designed for human consumption include the grain-eating European house sparrow *Passer domesticus* and the even more infamous African red-billed quelea (also called weavers) *Quelea quelea*, whose flocks may number millions and do damage similar to that caused by locusts but on a smaller scale. In North America the red-winged blackbird *Agelaius phoenicus* can also be very destructive to cereal crops. The droppings of the starling often destroy large areas of coniferous trees in which they roost, often in many thousands.

Complaints have also been levelled by farmers whose fields are visited by flocks of wildfowl. In the news in Britain of recent years has been the pink-footed goose *Anser brachyrhynchus* accused of destroying carrots in the region of the Wildfowl Trust's reserve at Martin Mere in Lancashire, and the Brent goose *Branta bernicla*, equally feared by the farmers of the south-east. The balanced truth of the matter seems to be that although some slight local damage may be done (important, however, to the particular local farmer who suffers) most of the crops do recover. Damage to crops by birds should not all be written off as bad luck. A little ecological forethought as to which crop should be planted where could save immensely on subsequent protection.

Anglers complain about such birds as cormorants *Phalacrocorax carbo*,

**Fig 114** Brent geese *Branta bernicla*

red-breasted mergansers *Mergus serrator*, goosanders *Mergus merganser*, herons *Ardea cinerea*, and even kingfishers *Alcedo atthis* depleting fish stocks, but the populations of these birds is never large enough to make all that much difference and in any case they may well be catching the slow-moving fish which are less healthy than the rest and therefore the net result may be actually to improve the swim. In the case of the heron the main item in its diet tends to be eels which in turn eat the eggs and fry of game fish. The heron in this case may be of positive benefit to the fishing but despite this it is still persecuted. A recent RSPB report suggested an alternative to this persecution. Research biologist Julie Meyer spent two years studying the problem of heron predation on fishfarms, and confirmed that shooting the birds is, in any case, no way out of the problem. As soon as one bird is shot another moves in, and greater losses for the farmer may result in disease, oxygen deprivation and poaching. Caging the pools is the ideal way to keep the herons out, but is expensive. The cheap and simple alternative is the humble ball and twine. Two strands of cord placed one above the other round pools, at heights of 25–35 cm (10–14 in) successfully prevent marauding birds from reaching their prey. As efficient is a chain of white polythene floats placed 30 cm (12 in) apart around the sides of pools or tanks.

Sea-based birds have also fallen foul of the shell fishing industry, and eiders *Somateria mollissima* have been accused of 'devastating' mussel beds and culls have been demanded. During 1974 argument raged around the part played by the oystercatcher *Haematopus ostralegus* in the demise of the cockle *Cerastoderma edule* industry centred on Burry Inlet lying between Glamorgan and Carmarthen. Prater (1974) put the peak winter count of oystercatchers in this area at 20,000 birds. A hasty decision to cut half the population over a two year period commencing in August 1975 was made by the Secretary of State for Wales and the Minister of Agriculture. This was carried out under the control of the South Wales Sea Fisheries Committee and a bounty of 25 pence per bird killed was paid. Prater calculated that 10 per cent of the British and 4 per cent of the West European and African population winter in this area, and that any culling in the future should only be allowed after careful study by competent scientific observers, and his words of wisdom do seem to have been heeded.

Before leaving the question of pests, the skylark *Alauda arvensis* and the moorhen *Gallinula chloropus*, both most unlikely suspects, must be mentioned. Skylarks eat the seedlings of early lettuce, peas and sugar beet, and moorhens often feed greedily on commercially grown violets. These are, however, relatively localised problems compared with species discussed earlier in this section.

## BIRDS AS VECTORS OF DISEASE

There are very few species of bird which pose a real threat to our health but the feral pigeon *Columba livia*, a descendent of the wild rock dove, most certainly does. It finds ledges on buildings and statues very similar to its native habitat on steep cliffs and sea stacks. The droppings of the huge congregations of birds have been found to contain many unpleasant things including a fungus called *Cryptococcus neoformans* which usually attacks only our skin, but can also damage the lungs and the nervous system. In extreme cases it may cause a type of meningitis which can prove fatal.

Disease can also be spread among those people handling pigeons. The disease, often called psittacosis, was particularly virile in Chicago during the winter of 1944/5 and subsequent research revealed that 45 per cent of pigeons were carriers of the disease. In Paris 66 per cent of the pigeons are infected, and this figure is exceeded in the pigeons of many British cities especially London and Manchester, but the disease seldom reaches the human population perhaps due to some sort of immunity. This is just as well because psittacosis is a most unpleasant disease, often thought only to be transmitted by parrots, but poultry workers and those involved in cleaning public buildings soiled by pigeon excreta are occasionally affected. At one time the disease was thought to be due to a virus but an intracellular parasite called *Chlamydia* is now known to be the cause. Human symptoms of the disease are a high temperature alternating with fits of the shivers and aching limbs. If treatment is not given quickly it can lead to respiratory problems and other complications but correct diagnosis followed by treatment with tetracycline drugs quickly bring the disease under control. In view of the fact that other birds besides parrots can spread psittacosis the much more accurate name for it is ornithosis.

Wild birds can also be responsible for the transmission of the economically crippling Newcastle Disease, more often referred to as fowlpest. Actually the name fowlpest covers two similar diseases, Newcastle Disease and fowl plague, and in Britain the latter is much the rarer of the two. Both are caused by a virus and Newcastle Disease has been known since 1926, and since this time considerable numbers of poultry in many parts of the world have been killed by it. A significant outbreak was noted in British pheasants during 1963 and in a number of gamebirds in 1970 and again in 1973. According to the Game Conservancy Booklet No. 6 the symptoms are 'general incoordination, a staggering gait, rapid head turning and alternate raising and lowering, a crouched posture often accompanied by a "wry-neck" appearance, fluffed feathers and a great increase in the amount of water drunk. Feeding continued throughout the early stages of the disease and the birds usually died in good condition (sic). Diarrhoea, white and dark green in colour occurred in most cases. In poultry respiratory signs are more common although this does depend upon the strain of virus'.

Wild birds are able to spread fowlpest although hard and fast evidence is difficult to come by though it is certain that they act as vectors of diseases affecting man himself, and his animal food supply. It has also been established that birds can also spread plant diseases, the house sparrow *Passer domesticus* being responsible for the spread of Tomato Mosaic Virus.

BIRDS AND AIRCRAFT

In 1960 a Lockheed Electra aircraft took off from Boston Airport, U.S.A. and collided with a flock of starlings: all 62 people on board died in the crash. Two years later a Vicker's Viscount flying at a height of 2,000 m (6,000 ft) above Maryland, U.S.A. struck a flock of whistling swans *Cygnus columbianus*. The impact caused the tailplane to shatter and the aircraft went into an uncontrolled dive: again there were no survivors of the 17 people aboard. Apart from these two disasters there have been scores of bird strikes which have almost brought tragedy to all major airports of the world. In 1962, immediately after take-off from Turnhouse Airport, Edinburgh, a Vanguard with 62 people on board struck a flock of gulls and only great skill on the part of the crew brought the aircraft safely to land. On checking the plane the remains of 125 dead gulls were removed from the engines. Now that some aircraft carry over 400 passengers the presence of birds in the vicinity of airports must be carefully monitored. Some have suggested strengthening the aircraft, but a flying machine faces the same problems as a flying bird: increased weight produces decreased efficiency.

The only thing to do is to study the birds. Urban planning, for example, should take account of the fact that at least half the bird strikes are due to gulls, and so any municipal rubbish dump should always be sited as far away from the airport as possible. A careful study should also be made of any migration fly-lines which pass close to airline routes. The lesser snow goose *Anser caerulescens*, for example, migrates through the Winnipeg region of Canada, and in the spring of 1969 a flock of these geese caused serious damage to a Boeing airliner. A detailed study of goose movement was then ordered. By 1974 Blokpoel had mapped the whole complex migratory movements, and although there is never any real chance of diverting the flocks of birds radar scans based on prior knowledge of movements will enable the aircraft to be directed away from the flocks. Such studies are now becoming standard around most of the world's major airports.

Airport planners must also face the fact that some species of bird actually find short grass ideal for feeding and roosting especially rooks *Corvus frugilegus* and starlings *Sturnus vulgaris*. Attempts to solve this problem have included allowing grass to grow longer than 20 cm (8 in) which the birds find less attractive as feeding habitat. This was tried at Vancouver to reduce the populations of passerines, especially starlings, but all it did was to encourage the rodent population to rise and this attracted large numbers of short-eared owls *Asio flammeus* which were every bit as much a menace as the grass-feeding birds had been.

It is during take-off and landing that the hazards are greatest, and so varied attempts have been made to scare the birds away from runway areas. London's Heathrow Airport has made use of a most unusual bird-scaring device. If a man stands 100 metres or so from a flock of birds and raises and lowers his arms 24 times a minute it is thought to simulate the wing beats of an eagle, and some ingrained instinct in the birds puts them to flight. Other airports have improved on this technique by flying remote controlled models built in the shape of hawks.

The production of loud noises has also been tried, but after a while the birds just get used to the noise and ignore it. Playing tape recordings of distress calls has also worked on occasions but seems to have little effect on either woodpigeons *Columba palumbus* or oystercatchers *Haematopus ostralegus*.

There is little which can be done to alter the situation at existing airports, but known ornithological data should certainly be fed into the early stages in the planning of new airports. What is more usual, however, is for the site to be selected on other more pressing economic grounds, later allocating the necessary finance to assess the risks of bird strikes and to devise methods of overcoming them.

## The Compromise between Birds and Man

In order to understand how birds and man interact, for good or evil, it is essential to be able to compare avian populations from one year to the next, and to know the length of the normal life span of a particular species. This is possible by reference to ringing returns (*see* Chapter 9). Any changes or ill effects can then be determined and any conservation measures decided upon the basis of these figures. Firstly let us look at how bird populations are counted, and then go on to consider briefly conservation, about which so much has been written in recent years.

### COUNTING BIRDS

Birds are very mobile creatures, often occurring in large numbers and spending a lot of time in thick cover or, in the case of pelagic birds, far out over the sea. Obviously, varying censusing techniques have been developed, including direct counts of seabird breeding 'cities' and inland colonial species, accurate counts of colourful or otherwise obvious species, and estimates of winter flocks by visual means. The British Trust for Ornithology (B.T.O.) has for over 20 years regularly censused common birds in certain areas and by far the great majority of these counts are carried out by amateur ornithologists.

**Fig 115** A gannetry, illustrating the problems of counting large numbers of birds

Seabird colonies are among the easiest, but also the most tedious, to count. Sailing around breeding ledges and actually doing a physical count will give reasonable results, but photographing the area, perhaps from the air, making an enlarged print and using a magnifying glass to count birds can be much more accurate since it effectively cancels out confusing movement. Computers are being used increasingly for this task. Another method of counting dense colonies, particularly gannets, terns and other species which tend to spread their nests evenly over suitable breeding habitat, is to calculate the total area occupied by the colony, then count a number of random areas and then multiply up, to arrive at the true population. Using several techniques of a like nature the Seabird Group in 1974 produced the figures in Table 13. When repeated at regular intervals these counts will show trends in populations which could be invaluable as pollution indicators.

Direct counts of inland species such as colonial nesting herons *Ardea cinerea* and rooks *Corvus frugilegus* were among the first to be attempted. Conspicuous species such as great crested grebes *Podiceps cristatus*, kingfishers *Alcedo atthis* and mute swans *Cygnus olor* also lend themselves easily to this technique, and expressed graphically the results show population

Table 13
SHOWING ESTIMATES OF THE NUMBERS OF SEABIRDS BREEDING ON THE COASTS OF BRITAIN AND IRELAND IN 1969–70

| Species | Thousands of Pairs |
|---|---|
| *Uria aalge*–guillemot | 577 |
| *Fratercula arctica*–puffin | 490 |
| *Rissa tridactyla*–kittiwake | 470 |
| *Larus argentatus*–herring gull | 333 |
| *Fulmarus glacialis*–fulmar | 306 |
| *Alca torda*–razorbill | 144 |
| *Sula bassana*–gannet | 138 |
| *Larus ridibundus*–black-headed gull | 74 |
| *Larus fuscus*–lesser black-backed gull | 47 |
| *Phalacrocorax aristotelis*–shag | 31 |
| *Sterna paradisaea*–arctic tern | 31 |
| *Larus marinus*–great black-backed gull | 22 |
| *Sterna hirundo*–common tern | 14 |
| *Larus canus*–common gull | 12 |
| *Sterna sandvicensis*–sandwich tern | 12 |
| *Cepphus grylle*–black guillemot | 8.3 |
| *Phalacrocorax carbo*–cormorant | 8.1 |
| *Stercorarius skua*–great skua | 3.1 |
| *Sterna dougallii*–roseate tern | 2.3 |
| *Sterna albifrons*–little tern | 1.8 |
| *Stercorarius parasiticus*–arctic skua | 1.1 |

*from Cramp, Bourne and Sanders, 1974,* The Seabirds of Britain and Ireland, *Collins, London*

trends very easily indeed. Such censuses of individual species can be organised on a national basis under the control of the British Trust for Ornithology or the Audubon Society, but local censuses can also be invaluable, especially when large scale construction projects are being planned. Winter counts of wildfowl and waders have regularly been conducted by the Wildfowl Trust and the British Trust for Ornithology respectively. Wildfowl on as many waters as possible are counted on fixed dates from September to April, each observer being responsible for his own home water. The counts are sent to the organising officer at the Wildfowl Trust who collates them, and the results show where the various species are found, what their populations are, and when the peak numbers occur. This gives a good idea of which watercourses the wildfowl prefer and the data are readily available to those involved in the planning of what are now called 'multiple uses of water resources'; when space has been allocated to picnickers, boaters, anglers and occasionally to shoots the figures often ensure a place for birds.

The British Trust for Ornithology's Waders and Estuaries Survey initiated by Tony Prater counts waders in a similar manner, but the practicalities are rather more difficult because waders are much smaller and are generally much more numerous and mobile. These factors mean that the counts are more likely to show errors, but two methods have been developed to improve the accuracy. Aerial photographs have been used, and also the ringers (banders) have been of great assistance in capturing birds and marking them with a harmless dye, usually on the underparts which are paler, or the tail coverts. This method is applicable only to a few species, mainly waders, as well as swans and geese. It is then possible to estimate the population by what has been called the capture/recapture method. A number of birds are caught, marked, and released. Later a second sample is taken and the number of recaptured individuals is noted. The population can be calculated from the formula

$$\text{Population} = \frac{\text{Number of birds in 1st sample} \times \text{number of birds in 2nd sample}}{\text{Number of marked individuals recaptured.}}$$

This method has its disadvantages, including the fact that random sampling is seldom complete and may be influenced by births, deaths or movements into or out of the area. Some individuals also tend to become trap-shy or trap-addicted so that the population of marked birds found in the second sample may be quite misleading. A reasonable estimate, however, is better than no estimate at all.

All that the statisticians amongst us now require is a method of censusing the smaller more secretive birds which spend much of their lives concealed in woodland and wasteland, field and fell. The Common Bird Census has proved invaluable in providing indications of population trends in such species; in Britain this again comes under the administrative umbrella of the

B.T.O. but many other countries also have similar schemes. As with any other census one set of results is worthless, and it is only when subsequent years can be compared that the true value can be appreciated. An individual's study of a small wood or area of farmland will also usually be of no real consequence, but if a hundred or more such areas have been censused and co-ordinated, data of vital importance is obtained. This dictates that the methods used to census the area should be rigidly laid down and stuck to from one year to the next, since any scientific exercise requires reliable and consistent method. At the beginning of the survey a large scale map of the area is obtained, and a number of tracings are made of this master. A preliminary visit is then made to the census area, and every geographical and botanical feature is marked on it, and subsequent regular visits are made to the plot. Each visit is recorded on a separate map and each singing bird (or any other breeding behaviour) is indicated using a definite set of symbols. A singing blackbird *Turdus merula* is indicated by the symbol (B), song thrush *Turdus philomelos* (ST), skylark *Alauda arvensis* (S) and so on, the symbol list being provided by the B.T.O. If the nest is found the symbol given is B*, ST* or S*; should the bird give an alarm call the symbol becomes B̲; two males fighting becomes BB, and so on. At the end of the breeding season a separate map can be plotted to show the breeding population for each species in the area and sent to the B.T.O. for processing and incorporation with other data. Gradually any trends will be indicated especially if the data is expressed graphically (*see* Fig. 116).

Any trend, such as the downward trend noted in the whitethroat *Sylvia communis*, can quickly be investigated. In the case of the whitethroat, 1968 was regarded as a good breeding year and yet the counts were dramatically down in 1969. Three things could have happened, namely, heavy losses on the autumn journey out to the wintering grounds; a catastrophe in this area during their stay; or tragedy on the way back to the breeding plots. Many suggestions were made including the effects of highly toxic chemicals used in sprays to kill locusts in Morocco but most of this had been done in the Sous valley in Morocco in November and December of 1968 long after the whitethroats should have crossed the Sahara and been safe in winter quarters. Local freak weather conditions were also discounted, but long term shortage of rainfall seems to have been the cause since the migrant birds do in the short term require water more than food. The Sahel Zone through which they pass had a higher than average rainfall during the 1950s, but in 1968 the rain failed to arrive and caused great human suffering in the region, crops failing, resulting in fewer insects and therefore fewer whitethroats. Many people died in the area and the whitethroat graph probably also reflects the human demise in the Sahel. This story was unravelled by the painstaking researches of P. Berthold (1973) and Winstanley *et al* (1974).

The Common Bird Census will also reveal upward trends, and nowhere has this been more dramatically shown than in the case of the collared dove

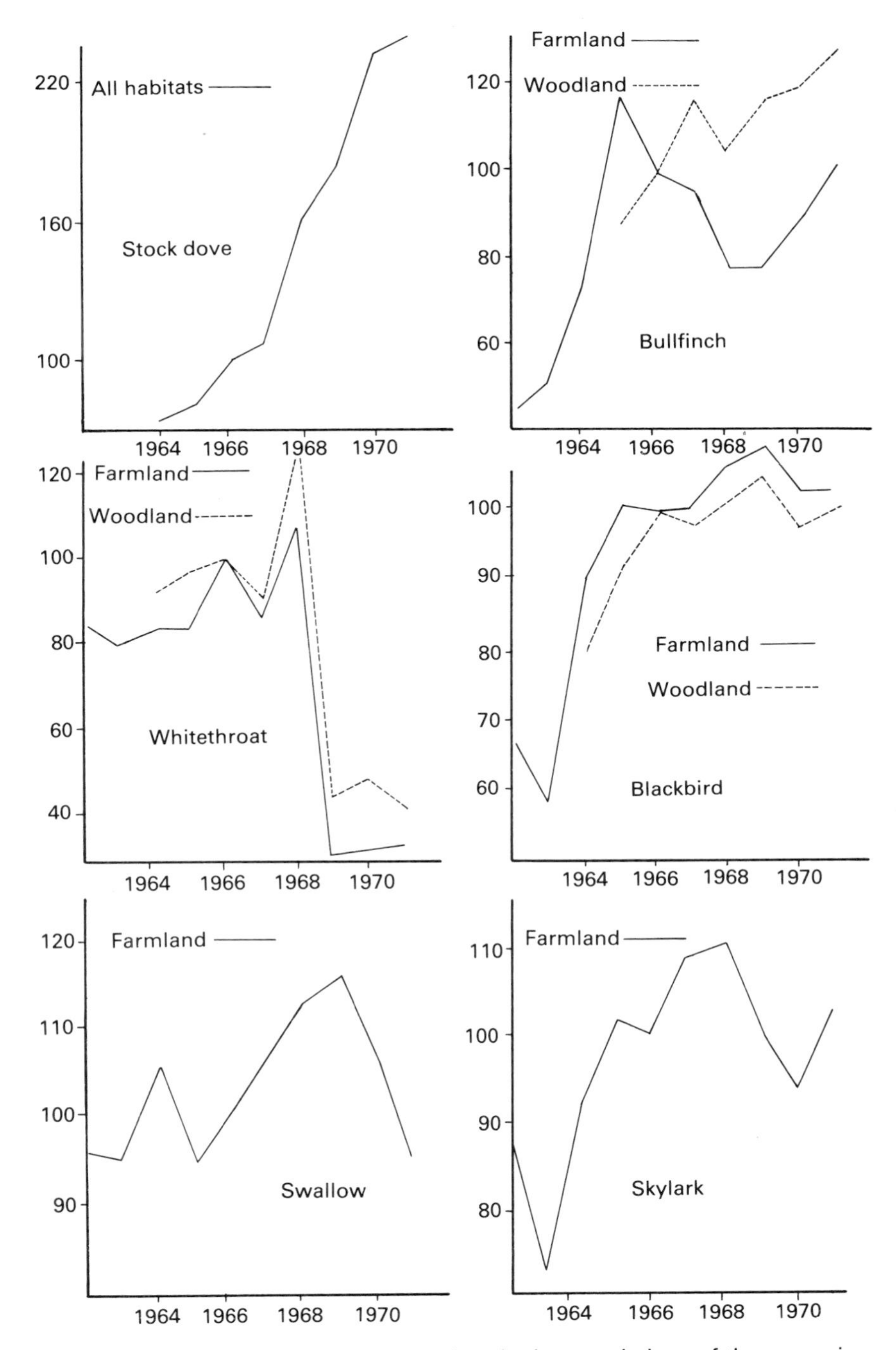


**Fig 116** The graphs show the fluctuations in the populations of these species over ten years. The figure 100 represents the population in 1964 and all other figures are a percentage of this

*Streptopelia decaocto*. Future work will be important in deciding upon whether the population levels are still increasing and will enable a decision to be made as to whether a selective cull is needed. Such censusing methods are essential tools of the conservationist, and it is this emotive subject that the final section of this book must mention.

## CONSERVATION

If a competition was organised to discover the most fashionable word of the past decade, then conservation must be one of the front runners. It is what may be described as an umbrella word, and sheltering beneath its widening canopy are those who wish to leave everything in the world just as it is. These people, however, usually have enough and some to spare. Many naturalists agree that this approach does not go far enough, does not acknowledge enough of the problems, to be much of a solution to the complex issues involved. Four aspects of bird conservation seem to be important at the moment.

Firstly, any valuable habitat remaining must be preserved if at all possible. In many cases, especially wetlands, there may be so little habitat remaining that nature must be prevented from reducing this still further by even normal vegetational succession where open water is replaced first by reeds, then by alder and finally becomes oak woodland. If we are to maintain a small place for waterbirds a nature reserve must first be set up and then managed. This ensures suitable habitat for birds such as the bittern *Botaurus stellaris* and the bearded reedling *Panurus biarmicus*.

Secondly, it must be admitted that we do need industrial development, and so ornithologists should not polarise themselves in an anti-industrial cloister, but look for compromise and methods of bringing pressure to bear in the most influential places. Industry can often be pursuaded to plough back some of their profits into extensive and impressive restoration programmes following, or even during, their enterprises.

Thirdly, existing derelict land can often be put to good use by creating habitats suitable for a given species or group of species. Nowhere is this more obvious than in parts of Britain when old gravel workings have been planted, and controlled flooding has produced ideal wader and wildfowl feeding and breeding habitat and handy roosts for hirundines and passerines. The aesthetic impact upon the local human residents is no small factor either.

Fourthly, it must be admitted that disruptions already created by human action, both deliberate and accidental, have driven some birds so far along the road to extinction that mere preservation of existing habitat and creation of additional habitat would be insufficient in themselves. What is sometimes required is a deliberate cold-blooded cull of their predators or competitors, however distasteful this may appear to be. On occasion even this will not be sufficient and so breeding the species in captivity and then releasing part of the captive population into ideal habitats and closely monitoring the in-

**Fig 117** The ne-ne *Branta sandvicensis*, an example of a severely endangered species

dividuals may have to be attempted. This has proved of some value in the case of the ne-ne *Branta sandvicensis*, and also the sea eagle *Haliaeetus albicilla*, though they are very different cases, the former being the more urgent as the species faces total extinction while the latter is merely a reintroduction from areas of abundance.

The value of birds as environmental indicators is now fully appreciated by more ornithologists, but however much populations are studied and analysed no real progress can be made until the facts are understood and accepted by the large majority of the public. Education has a vital role and bird societies and publishers of books and periodicals all have their part to play in this essential enterprise. The past history of birds is a long one; their present position is one of armed truce, but their future does seem brighter than was the case a century ago. Let us all hope for the sake of future generations that I am not mistaken.

# Bibliography


## General Works of Reference

DORST, J. (1974) *The Life of Birds*, 2 vols. Weidenfeld and Nicolson, London.

FISHER, J. AND FLEGG, J. (1974) *Watching Birds*, Poyser, Staffordshire.

MIVART, ST. GEORGE (1892) *Birds, the Elements of Ornithology*, Porter and Dalan, London.

NICHOLSON, E.M. (1927) *How Birds Live*, Williams and Norgate, London.

RAND, A.L. (1974) *Ornithology*, Pelican, London.

ROBERTS, M.B.V. (1976) *Biology: a Functional Approach*, Nelson, Sunbury-on-Thames.

THOMSON, A.L. (1964) *A New Dictionary of Birds*, Nelson, Sunbury-on-Thames.

TNYE AND BERGER (2nd Ed. 1976) *Fundamentals of Ornithology*, John Wiley, Chichester.

WELTY, J.C. (2nd Ed. 1975) *The Life of Birds*, Saunders, Eastbourne.

YAPP, W.B. (1970) *The Life and Organisation of Birds*, Edward Arnold, London.

## Chapter 1

ATTENBOROUGH, D. (1978) *Life on Earth*, BBC/Collins, London.

BRODKORB, P. (1963) *Catalogue of Fossil Birds* 2 vols. Florida State Museum.

BUICK, T.L. (1932) *The Mystery of the Moa*, Francis Edwards, London.

CHARIG, A. (1979) *A New Look at Dinosaurs*, Heinemann, London.

GRIEVE, S. (1885) *The Great Auk or Garefowl*, T.C. Jack, London.

HACHISKUA, M. (1953) *The Dodo and Kindred Birds*, Witherby, London.

HARRIS, M. (1978) *A Field Guide to the Birds of the Galapagos*, Collins, London.

HARRISON, C.J.O. (1979) Additional birds from the Lower Pleistocene of Olduvai, Tanzania and Potential Evidence of Pleistocene Bird Migration, *Ibis* 122, pp. 530–532.

HARRISON, C.J.O. (1982) *Fossil Birds from Africa South of the Sahara, in the Collection of the British Museum (Natural History)*, British Museum (Natural History), London.

LACK, D. (1947) *Darwin's Finches*, Cambridge University Press.

LACK, D. (1974) *Evolution Illustrated by Waterfowl*, Blackwell, Oxford.

LOWE, P. (1933) On Some Struthions Remains 1. Descriptions of some Pelvic Remains of a Large Fossil Ostrich *Struthio oldowayi* from the Lower Pleistocene of Oldawayi (Tanganyika Territory) *Ibis* 13/3, pp. 652–658.

MAYR, E. (1963) *Animal Species and Evolution*, Harvard University Press.

PARKIN, T. (1911) *The Great Auk. A Record of Sales of Birds and Eggs by Public Auction in Great Britain 1806–1910*, Burfield and Pennels, London.

RALLING, C. (ed) (1978) *The Voyage of Charles Darwin*, BBC, London.

ROMER, A. (1959) *The Vertebrate Story*, Chicago University Press.

SWINTON, W.E. (1975) *Fossil Birds*, British Museum (Natural History), London.

## Chapter 2


### *General References*

GRUSON, E.S. (1976) *Check List of Birds of the World*, Collins, London.

SEEBOHM, H. (1890) *The Classification of Birds*, R.H. Porter, London.

TATE, P. (1979) *A Century of Bird Books*, Witherby, London.

### *Order 1*

HAVERSCHMIDT, W. (1968) *Birds of Surinam*, Edinburgh University Press.

HUDSON, W.H. (1920) *Birds of La Plata* Vol 2, London University Press.

RIDGELY, H. (1976) *A Guide to the Birds of Panama*, Princeton University Press.

WETMORE, A. (1965) *The Birds of the Republic of Panama*, Vol 1, Smithson Miscellaneous Collection pp. 5–24, Washington, D.C.

### *Order 2*

JACKSON, F.J. (1938) *The Birds of Kenya Colony and the Uganda Protectorate*, Vol 1, London University Press.

SAUER, E.G.F. and Sauer, E.M. (1966) The Behaviour and Ecology of the South African Ostrich *The Living Bird* pp. 45–75.

### *Order 3*

BRUNING, D.F. (1974) Social Structure and Reproductive Behaviour of the Greater Rhea, *The Living Bird* pp. 251–294.

HUDSON, W.H. (1920) *Birds of La Plata*, Gurney, London.

### *Order 4*

CAMPBELL, A.J. (1901) *Nests and Eggs of Australian Birds Vol II*, Sheffield University Press.

CAYLEY, N. (1966) *What Bird is That?* Sydney University Press.

FRITH, H.J. (ed) (1969) *Birds in the Australian High Country*, Sydney University Press.

RAND, A.L. AND GILLIARD, E.T. (1967) *Handbook of New Guinea Birds*, New York University Press.

### *Order 5*

ANDERSON, R.J. (1930) *New Zealand Birds*, Wellington Press, New Zealand.

BULLER, W.L. (1888) *A History of the Birds of New Zealand Vol 2*, London University Press.

CALDER, W.A. (1911) The Kiwi and Egg Design: Evolution as a Package Deal, *Biological Science* 29, pp. 461–467.

OLIVER, W.R.B. (1930) *New Zealand Birds*, Wellington Press, New Zealand.

### *Order 6*

GODFRY, W.E. (1966) *Birds of Canada*, Ottowa University Press.

HARVIE BROWN, J.A. (1895) *Atlas of Scotland*, Oliver and Boyd, Edinburgh.

RANKIN, N. (1947) *Haunts of British Divers*, Collins, London.

TATE, D.J. and TATE, (Jnr.) J. (1970) Mating Behaviour of the Common Loon *Auk* 87, pp. 125–30.

## *Order 7*

HARRISON, T.H. AND HOLLOM, P.A.D. (1932) The Great Crested Crebe Enquiry 1931, *British Birds* 26.

HUXLEY, J.S. (1914) The Courtship Habits of the Great Crested Grebe, *Proceedings of the Zoological Society of London* pp. 491–562.

MUNRO, J.A. (1941) Studies of Waterfowl in British Columbia: The Grebes *British Columbia Museum Occasional Paper* Number 3.

PRESTT, I. AND MILLS, D.H. (1966) A Census of Great Crested Grebe in Britiain, 1965, *Bird Study* 13.

STORER, R.W. (1963) Observations on the Great Grebe *Condor*, 65, pp. 279–288.

## *Order 8*

AUSTIN, O.L. JR. (1968) Antarctic Bird Studies, *Antarctic Research Series 12*, American Geophysical Union.

KEARTON, C. (1930) *The Island of Penguins*, Longmans, London.

PETTINGILL, O.S. JR. (1964) Penguins Ashore at the Falkland Islands *Living Bird*, pp. 45–64.

RICHDALE, L.E. (1951) *Sexual Behaviour in Penguins*, Lawrence, Kansas University Press.

ROBERTS, R. (ed) (1967) *Edward Wilson's Birds of the Antarctic*, Blandford Press, Poole.

SIMPSON, G.G. (1976) *Penguins Past and Present, Here and There*, Yale University Press.

SPARKS AND SOPER (1967) *Penguins*, David and Charles, Newton Abbot.

WHITLOCK, R. (1977) *Penguins*, Wayland, Hove.

## *Order 9*

CRAMP, BOURNE AND SAUNDERS (1974) *The Seabirds of Britain and Ireland*, Collins, London.

FISHER, J. (1952) *The Fulmar*, Collins, London.

FISHER, J. (1966) *The Fulmar Population of Britain and Ireland, 1959 Bird Study* 13, pp.5–76.

FISHER, J. AND LOCKLEY, R.M. (1954) *Seabirds*, Collins, London.

LOCKLEY, R.M. (1942) *Shearwaters*, Collins, London.

PALMER, R.S. (ed) (1962) *Handbook of North American Birds*, Vol I, New Haven.

WILLIAMSON, K. AND BOYD, J.M. (1960) *St. Kilda Summer*, Collins, London.

## *Order 10*

FISHER AND VEVERS, G. (1943/1944) *The Breeding Distribution, History and Population of the North Atlantic Gannet* Sula bassana *Journal of Animal Ecology* 12, pp. 173–213; 13 pp. 49–62.

FRITH, H.J. (1969) *Birds in the Australian High Country*, Sydney University Press.

NELSON, B. (1978) *The Gannet*, Poyser, Staffordshire.

NELSON, B. (1978) *The Sulidae: Gannets and Boobies*, Aberdeen University Press.

OLIVER, W.R.B. (1955) *Birds of New Zealand*, Wellington Press, New Zealand.

PALMER, R.H. (ed) (1962) *Handbook of North American Birds*, Vol I, New Haven.

### *Order 11*

BROWN, L. (1959) *The Mystery of Flamingoes*, Country Life, Feltham.

KEAR, J. AND DUPLAIR-HALL, N. (eds) (1975) *The Flamingoes*, Poyser, Staffordshire.

LOWE, F.A. (1954) *The Heron*, Collins, London.

MCVAUGH, JR., W. (1972) The Development of Four North American Herons, *Living Bird* 155–137.

MEYERRIECKS, A.J. (1966) Comparative Breeding Behaviour of Four Species of North American Herons, *Publication of the Nuttal Ornithology Club* No. 2.

### *Order 12*

ATKINSON-WILLES, G.L. (1963) *Wildfowl in Great Britain*, H.M.S.O. London.

DELACOUR, J. (1975 6th imp.) *Waterfowl of the World*, 4 Vols, Country Life, Feltham.

OGILVIE, M.A. (1975) *Ducks of Britain and Europe*, Poyser, Staffordshire.

OWEN, M. (1977) *Wildfowl of Europe*, Macmillan, London.

SCOTT, P. (1972) *The Swans*, Michael Joseph, London.

SOOTHILL, E. AND WHITEHEAD, P. (1978) *Wildfowl of the World*, Blandford Press, Poole.

TODD, F.S. (1979) *Waterfowl: Ducks, Geese and Swans of the World*, California University Press.

### *Order 13*

BROWN, L. (1976) *British Birds of Prey*, Collins, London.

BROWN, L. (1976) *Birds of Prey, Their Biology and Ecology*, Hamlyn, Feltham.

EVERETT, M. (1977) *The Golden Eagle*, Blackwood, Edinburgh.

GLASIER, P. (1978) *Falconry and Hawking*, Batsford, London.

HARRIS, J.T. (1979) *The Peregrine Falcon in Greenland*, University of Missouri.

PORTER, R.F., WILLIS, I., CHRISTENSEN, S., NIELSEN, B.P. (1976) *Flight Identification of European Raptors*, Poyser, Staffordshire.

RATCLIFFE, D. (1980) *The Peregrine Falcon*, Poyser, Staffordshire.

WATSON, D. (1977) *The Hen Harrier*, Poyser, Staffordshire.

WOODFORD, M.H. (3rd Ed. 1977) *A Manual of Falconry*, A. and C. Black, London.

### *Order 14*

BAXTER, E.V. AND RINTOUL, L.J. (1953) *The Birds of Scotland* Vol 2, Oliver and Boyd, Edinburgh.

DELACOUR, J. (1951) *The Pheasants of the World*, Country Life, Feltham.

LACK, D. (1939) The Display of the Blackcock, *British Birds* 32, pp. 290–303.

MOSS, R. (1980) Why are Capercaillie Cocks so big? *British Birds* 73, pp. 440–447.

PALMAR, C.E. (1971) *Blackgame*, Forestry Commission leaflet No. 66.

PALMAR, C.E. (1976) *The Capercaillie*, Forestry Commission Forest Record 109.

PINOWSKI, J. AND KENDEIGH, S.C. (ed.) (1977) *Granivorous Birds in Ecosystems*, Cambridge University Press.

VESEY-FITZGERALD, B. (1946) *British Game*, Collins, London.

WAYRE, P. (1969) *A Guide to the Pheasants of the World*, SPC, London.

### *Order 15*

HOWARD, E. (1940) *A Waterhen's World*, Cambridge University Press.

MCNULTY, F. (1967) *The Whooping Crane*, Longman Green, London.
RIPLEY, S.D. (1977) *Rails of the World*, Boston University Press.

### *Order 16*

ARNOLD, A. (1924) *British Waders*, Cambridge University Press.
HALE, W.G. (1980) *Waders*, Collins, London.
NETHERSOLE-THOMPSON, D. (1951) *The Greenshank*, Collins, London.
NETHERSOLE-THOMPSON, D. (1980) *Greenshanks*, Poyser, Staffordshire.
PRATER, A.J. (1981) *Estuary Birds of Britain and Ireland*, Poyser, Staffordshire.
PRATER, MARCHANT VUORINERI (1977) Guide No. 17, British Trust For Ornithology, Tring.
RATCLIFFE, D. (1976) *Observations on the Breeding of the Golden Plover in the British Isles*, British Trust for Ornithology, Tring.
SEIGNE, J.W. AND KEITH, E.C. (1936) *Woodcock and Snipe*, Philip Allan, Oxford.
TINBERGEN, N. (1953) *The Herring Gull's World*, Collins, London.
VAUGHAN, R. (1980) *Plovers*, Terence Dalton, Sudbury.

### *Order 17*

GOODWIN, D. (1977) *Pigeons and Doves of the World*, Cornell University Press.
MCNEILLIE, A. (1977) *Guide to Pigeons of the World*, Elsevier-Phaidon, Oxford.
MURTON, R.K. (1965) *The Wood Pigeon*, Collins, London.
SIMMS, E. (1979) *The Public Life of the Street Pigeon*, Hutchinson, London.

### *Order 18*

BAKER, E.C.S. (1942) *Cuckoo Problems*, Witherby, London.
CHANCE, E.P. (1922) *The Cuckoo's Secret*, Sidgwick and Jackson, London.
CHANCE, E.P. (1940) *The Truth about the Cuckoo*, Country Life, Feltham.
JAPP, A.H. (1899) *Our Common Cuckoo and other Cuckoos and Parasitical Birds*, Burough, London.
KNIGHT, MARWELL (1955) *A Cuckoo in the House*, Methuen, London.
WYLLIE, I. (1981) *The Cuckoo*, Batsford, london.

### *Order 19*

EASTMAN, JR. W.R. AND HUNT, A.C. (1966) *The Parrots of Australia*, Sydney University Press.
FORSHAW, JIM (1973) *Parrots of the World*, Doubleday, New York.
HAYWARD, J. (1979) *Lovebirds and their Colour Mutations*, Blandford Press, Poole.
LOW, R. (1980) *Parrots, Their Care and Breeding*, Blandford Press, Poole.
SELK-SMITH, D. (1903) *Parrakeets*, Gronvold, New York.

### *Order 20*

CHAPIN, J.P. (1963) The Touracos: an African Bird Family, *Living Bird* 57–68.
JACKSON, F.J. (1938) *The Birds of Kenya Colony and Uganda Protectorate*, Vol I, Witherby, London.
MACKWORTH-PRAED, C.W. AND GRANT, C.B.H. (1962) *Birds of the Southern Third of Africa*, Witherby, London.

### *Order 21*

CATCHPOLE, C. (1976) *Owls*, Bodley Head, London.

EVERETT, M. (1977) *A Natural History of Owls*, Hamlyn, Feltham.
FREETHY, RON (1976) Little Owl Flying at Dunlin, *British Birds* 69, p. 272.
HIBBERT-WARE, A. (1938) *The Little Owl, An Examination of its Food Habits*, Witherby, London.
SPARKS AND SOPER (1970) *Owls, Their History and Natural History*, David & Charles, Newton Abbot.
YALDEN, D.W. (1977) *The Identification of Remains in Owl Pellets*, Mammal Society, Reading.

### *Order 22*

ALI, S. AND RIPLEY, S.D. (1970) *Handbook of the Birds of India and Pakistan* Vol. 4, Bombay University Press.
MEES G.F. (1977) *Geographical variation of Caprimulgus macrurus* Horsfield (Aves, Caprimulgidae) Zool. Verh. 155.

### *Order 23*

CALDER W.A. (1975) Daylength and the Hummingbirds Use of Time *Auk* 92, pp. 81–97.
LACK, D. (1956) *Swifts in a Tower*, Methuen, London.
MCATEE W.L. (1947) *Torpidity in Birds*, 38, pp. 191–206.
SCHEITHAUER, W. (1967) *Hummingbirds*, Arthur Barker, London.
SKUTCH, A.F. (1974) *The Life of the Hummingbird*, Octopus, London.
STILES, F.G. (1980) The Annual Cycle in a Tropical Wet Forest Hummingbird Community, *Ibis* 122, p. 322.
WORTH, C.B. (1943) Notes on the Chimney Swift, *Auk* 60, pp. 558–664.

### *Order 24*

CADE, T.J. AND GREENWALD, L.I. (1966) Drinking Behaviour of Mousebirds in the Mambi Desert, *Auk* 83, pp. 126–28.
MACKWORTH PRAED, C.N. AND GRANT, C.H.B., (1962) *Birds of the Southern Third of Africa*, Witherby, London.

### *Order 25*

ALI, S. AND RIPLEY, S.D. (1970) *Handbook of the Birds of India and Pakistan* Vol. 4, p. 60ff, Bombay University Press.
SKUTCH, A.F. (1944) Life History of the Quetzel, *Condor* 46, pp. 213–235.
WETMORE, A. (1968) The Birds of the Republic of Panama, 2, *Smithson Miscellaneous Collection* 150, pp. 379–419.

### *Order 26*

ALI, S. AND RIPLEY, S.D. (1970) *Handbook of the Birds of India and Pakistan*, Bombay University Press.
BOAG, D. (1982) *The Kingfisher*, Blandford Press, Poole.
CLANCEY, P.A. (1964) *The Birds of Natal and Zululand*, Oliver and Boyd, Edinburgh.
DEMENTEV, G.P. (1966) *Birds of the Soviet Union*, Vol. 1, Washington University Press.
EASTMAN, R. (1969) *The Kingfisher*, Collins, London.

FRITH, H.J. (ed.) (1969) *Birds in the Australian High Country*, Sydney University Press.

FRY, C.H. (1979) The Origin of Afrotropical Kingfishers, *Ibis* 122, p. 57.

MEYER DE SCHAUENSEE, R. (1970) *A Guide to the Birds of South America*, Wyndewood, London.

SKEAD, C.J. (1950) A Study of the African Hoopoe, *Ibis* 92, pp. 434–463.

*Order 27*

LEUTSCHER, A. (1973) *Woodpeckers*, Franklin Watts, London.

PALMAR, C.E. (1974) *Woodpeckers in Woodlands*, Forestry Commission Record, No. 92.

SIELMANN, H. (1960) *My Year with the Woodpeckers*, Collins, London.

SKUTCH, A. (1969) Life Histories of Central American Birds, III, *Pacific Coast Avifauna*, No. 35.

SPRING, L.W. (1976) Climbing and Pecking Adaptations in some North American Woodpeckers, *Condor* 67, pp. 457–488.

*Order 28*

ANGELL, T. (1978) *Ravens, Crows, Magpies and Jays*, University of Washington Press.

ARMSTRONG, E.A. (1955) *The Wren*, Collins, London.

BARNES, J.A.G. (1975) *The Titmice of the British Isles*, David and Charles, Newton Abbot.

BROWN, P.E. AND DAVIES, M.G. (1949) *Reed Warblers*, Foy Publications, London.

CAMPBELL, B. (1974) *The Crested Tit*, Forestry Commission Record, No. 98.

DIXON, C. (1897) *Our Favourite Song Birds*, Lawrence and Bullen, London.

GOODWIN DEREK (1976) *Crows of the World*, British Museum (Natural History), London.

HILLSTEAD, A.F.C. (1945) *The Blackbird*, Faber and Faber, London.

HOSKING, E. AND NEWBERRY, C. (1946) *The Swallow*, Collins, London.

HOWARD, H. ELLIOT (1914), *The British Warblers*, 2 Vols, R.H. Parker, London.

INGRAM, C. (1974) *The Migration of the Swallow*, Witherby, London.

LACK, D. (1943) *The Life of the Robin*, Witherby, London.

MOUNTFORD, G. (1957) *The Hawfinch*, Collins, London.

NETHERSOLE-THOMPSON, D. (1966) *The Snow Bunting*, Oliver and Boyd, Edinburgh.

NETHERSOLE-THOMPSON, D. (1975) *Pine Crossbills*, Poyser, Berkhamstead.

NEWTON, IAN (1972) *Finches*, Collins, London.

PERRINS, C. (1978) *British Tits*, Collins, London.

PIKE, O.G. (1932) *The Nightingale, its story and its song*, Arrowsmith, London.

SHEEHAM, ANGELA (1976) *The Song Thrush*, Angus and Robertson, Brighton.

SIMMS, ERIC (1978) *British Thrushes*, Collins, London.

SMITH, STUART (1950) *The Yellow Wagtail*, Collins, London.

SOWERBY, J.G. (1895) *Rooks and their Neighbours*, Gay and Hancock, London.

SUMMERS-SMITH, J.D. (1963) *The House Sparrow*, Collins, London.

TICEHURST, C.B. (1938) *A Systematic Review of the Genus* Phylloscopus, British Museum (Natural History), London.

WILLIAMSON, KENNETH (1960–64) *On Warblers: Identification for Ringers*, B.T.O. Guides 7, 8 and 9, British Trust for Ornithology, Tring.
YEATES, G.K. (1934) *The Life of the Rook*, Allan, London.

## Chapter 3

BEEBE, C.W. (1907) *The Bird, its Form and Function*, Constable, London.
ELDER, W.H. (1954) 'The Oil Gland of Birds', *Wilson Bulletin* 66, 6–31.
KING, A.S. AND MCCLELLAND, J.M. (1975) *Outlines of Avian Anatomy*, Baillière Tindall, London.
MARSHALL, A.J. (1960) *Biology and Comparative Physiology of Birds*, 2 vols, Academic Press, London.
WORDEN, A.N. (1957) Functional Anatomy of Birds, *Cage Birds*, London.

## Chapter 4

AYMAR, G.C. (1936) *Bird Flight*, Allen Lane, London.
BERNSTEIN, H.M., THOMAS, S.P. AND SCHMIDT-NEILSON, K. (1973) Power Input during Flight of the Fish Crow, *Corvus ossifragus*, *Journal of Experimental Biology* 58, pp. 401–410.
GRAY, J. (1953) *How Animals Move*, Cambridge University Press.
KAUFMANN, J. (1970) *Birds in Flight*, World's Work, Tadworth.
PEARSON, O.P. (1964) Metabolism and Heat Loss during Flight in Pigeons, *Condor* 66, pp. 182–185.
PENNYCUICK, C.J. (1968a) A Wind Tunnel Study of Gliding Flight in the Pigeon, *Columba livia*, *Journal of Experimental Biology* 48, pp. 509–526.
PENNYCUICK, C.J. (1968b) Power Requirements for Horizontal Flight in the Pigeon *Columba livia*, *Journal of Experimental Biology* 49, pp. 527–555.
PENNYCUICK C.J. (1972) *Animal Flight*, Edward Arnold, London.
PRICE, NANCY (1947) *Wonder of Wings*, Gollancz, London.
PYCRAFT, W.P. (1922) *Birds in Flight*, Gay and Hancock, London.
SAVILE, D.B.O. (1950) The Flight Mechanism of Swifts and Hummingbirds, *Auk* 67, pp. 499–504.
SCHUCHMANN, KARL L. (1979) Metabolism of Flying Hummingbird, *Ibis* 121.
TUCKER, V.A. The Energetics of Bird Flight, *Scientific American*.
TUCKER, V.A. (1972) Metabolism During Flight in the Laughing Gull, *Larus atricilla* *American Journal of Physiology* 222, pp. 237–245.
URRY, D. AND K. (1970) *Flying Birds*, Vernon and Yates, Godalming.
WELTY, J.C. (1963) *The Life of Birds*, W.B. Saunders, Eastbourne.

## Chapter 5

BRACKENBURY, J.H. (1979) Aerocoustics of the vocal organ of birds, *Journal of Theoretical Biology* 81, pp. 341–349.
BRACKENBURY, J.H. (1979) Power Capabilities of the Avian Sound Producing System, *Journal of Experimental Biology* 78, pp. 163–166.
BRACKENBURY, J.H. (1980) Respiration and Production of Sound by Birds, *Biological Review* 55, pp. 363–378.

BRETZ, W.L. AND SCHMIDT-NIELSON, K. (1971) Bird Respiration: Flow Patterns in the Duck Lung, *Journal of Experimental Biology* 54/1.

FLYNN, R.K. AND GESSAMAN, J.A. (1979) An Evaluation of the Heart Rate as a Measure of Daily Metabolism in Pigeons, *Columba livia*, *Comparative Biochemistry and Physiology* 63A, pp. 511–514.

HUGHES, G.M. (1973) *The Vertebrate Lung*, Biology Readers No. 59, Oxford University Press.

KING, A.S. (1964) in Structural and Functional Aspects of the Avian Lungs and Air Sacs, part of *General and Experimental Zoology* Vol. II, ed. Felts and Harrison, Academic Press, London.

SCHMIDT-NIELSON, K. (Dec. 1971) How birds breathe, *Scientific American* 225/6, pp. 72–79.

TOMLINSON, J.T. (1963) The Breathing of Birds in Flight, *Condor* 65, pp. 514–516.

TUCKER, V.A. (Feb 1968) Respiratory Physiology of House Sparrows in relation to High Altitude Flight, *Journal of Experimental Biology* 48/1.

WORDEN, A.N. (1957) Functional Anatomy of Birds, Cage Birds, London.

## Chapter 6

BOCK, W.J. (1961) Salivary Glands in the Gray Jays, (*Perisoreus*), *Auk* 78, pp. 355–365.

BOCK. W.J. (1978) Tongue Morphology and Affinities of the Hawaiian Honeycreeper *Melamprosops phaeosoma*, *Ibis* 120, pp. 467–479.

BURTON, P.J.K. (1974) *Feeding and Feeding Apparatus in Waders*, British Museum (Natural History), London.

CADE, T.J. AND BARTHOLOMEW, G.A. (1959) Sea-water and salt utilization by Savannah Sparrow, *Physiolozical Zoology* 32, pp. 230–238.

COOCH, F.G. (1964) A Preliminary Study of the Survival Value of a Functional Salt Gland in Prairie Anatidae, *Auk* 81, pp. 380–393.

DUGAN, P.J., EVANS, P.R., GOODYER, O. AND DAVIDSON, N.C. (1981) Winter Fat Reserves in Shorebirds: Disturbance of Regulated Levels by Severe Weather Conditions, *Ibis* 123, pp. 359–363.

FREETHY, RON (1976) Little Owl Flying at Dunlin, *British Birds* 69, p. 272.

FREETHY, RON (1979) Kestrel Robbing Merlin, *British Birds* 72, pp. 336–37.

GREEN, J. (1968) *The Biology of Estuarine Animals*, Sidgwick and Jackson, London.

HAILS, C.J. AND AMIRRUDIN, A. (1981) Food Samples and Selectivity of White Bellied Swiftlets *Collocalia esculenta*, *Ibis* 123, pp. 328–333.

HILL, K.J. (1971) *Physiology and Biochemistry of the Fowl*, Vol. 1, Academic Press, London.

LINT, K.C. AND LINT, A.M. (1981) *Diets for Birds in Captivity*, Blandford Press, Poole.

NEWSTEAD, R. (1908) *The Food of Some British Birds*, Board of Agriculture, 1908.

OSBORNE, T.O. (1981) Ecology of the Red-Necked Falcon *Falco chicquera* in Zambia, *Ibis*, 123, pp. 289–297.

WORDEN, A.N. (1957) *The Functional Anatomy of Birds*, Cage Birds, London.

## Chapter 7

HARTMAN, F.A. AND ALBERTIN, R.H. (1951) A Preliminary study of the Avian

Adrenal, *Auk* 68, pp. 202–209.
MARSHALL, A.J. (ed.) (1961) *Biology and Comparative Physiology of Birds*, Williams, New York and London.
PEARSON, R. (1972) *The Avian Brain*, Academic Press, New York.
PUMPHREY, R.J. (1948) The Sense Organs of Birds, *Ibis* XC, pp. 171–179.
SAIFF, E.I. (1978) The Middle Ear of the Skull of Birds: the Pelicaniformes and Ciconiformes, *Zoological Journal of the Linnaean Society* 63, pp. 315–370.
VAN TIENHOVEN, A. (1969) *The Nervous System of Birds*, Poultry Science 48, pp. 10–16.
WORDEN, A.N. (1957) *Functional Anatomy of Birds*, Cage Birds, London.

## Chapter 8

ARMSTRONG, E.A. (1973) *A Study of Bird Song*, Oxford University Press.
ARMSTRONG, E.A. (1975) *Discovering Bird Song*, Shire Publications, Aylesbury.
CATCHPOLE, C. (1981) Why Do Birds Sing? New Scientist 90/1247, pp. 29–31.
DELAMAIN, J. (1928) *Why Birds Sing*, Gollancz, London.
DOWSETT-LEMAIRE, F. (1979) The Imitative Range of the Song of the Marsh Warbler *Acrocephalus palustris*, with Special Reference to Imitations of African Birds, *Ibis* 121, pp. 453–465.
FREETHY, RON (1980) Division of Labour between Dippers Building Nests, *British Birds* 73, p. 352.
GREENEWALT, C.H. (1968) *Bird Song: Acoustics and Physiology*, Smithsonian Institute Press, Washington, D.C.
GREENEWALT, C.H. (Nov. 1969) How Birds Sing, *Scientific American*.
HARTSHORNE, C. (1973) *Born to Sing: an Interpretation and World Survey of Bird Song*, Indiana University Press.
HINDE, R.A. (1972) *Non-verbal Communication*, Cambridge University Press.
JELLIS, R. (1977) *Bird Sounds and Their Meaning*, B.B.C. Publications, London.
KREBS, J. (1976) Bird Song and Territorial Defence, *New Scientist* 70, pp. 534–536.
MARSHALL, A.J. (1950) The Function of Vocal Mimicry, *Emu* 50, pp. 5–16.
MURTON, R.K. AND WESTWOOD, N.J. (1978) *Avian Breeding Cycles*, Clarendon Press, Oxford.
PALMER, S. AND BOSWALL, J. (eds 1972) *The Peterson Field Guide to the Bird Songs of Britain and Europe*, 2 stereo discs, RFLP 5007 and RFLP 5008. Stockholm Sveriges Radio Förlag 5, 105–110.
PAYNE, R.B. (1979) Song Structure, Behaviour and Sequence of Song Types in a Population of Village Indigobirds, *Animal Behaviour*, pp. 99–1013.
PICKSTOCK, J.C., KREBS, J.R. AND BRADBURY, S. (1980) Quantitative Comparison of Sonograms using an Automatic Image Analyser. Application to Song Dialects of Chaffinches, *Fringilla coelebs*, *Ibis* 122, pp. 103–109.
SIMON, H. (1977) *The Courtship of Birds*, Cassell, London.
THIELCKE, G. (1975) *Bird Sounds*, University of Michigan Press.
THORPE, W.H. (1958) The Learning of Song Patterns by Birds, with special reference to the song of the Chaffinch, *Ibis* 109, pp. 535–570.
THORPE, W.H. (1961) *Bird Song: the Biology of Vocal Communication and Expression in Birds*, Cambridge University Press.
WORDEN, A.N. (1957) Functional Anatomy of Birds, *Cage Birds*, London.

## Chapter 9

ABEL, K.P. (1977) The Orientation of Passerine Nocturnal Migrants Following Offshore Draft, *Auk* 94, pp. 320–330.

ALERSTAM, T. (1979) Wind as a Selective Agent in Bird Migration, *Scandinavian Ornithology* 10, pp. 76–95.

BOURNE, W.R.P. (1979) The Midnight Descent, Dawn Ascent and Re-Orientation of Land Birds Migrating across the North Sea in Autumn, *Ibis* 122, pp. 536–540.

BURTT, H.E. (1967) *The Psychology of Birds*, Macmillan, London.

DORST, J. (1962) *The Migrations of Birds*, Heinemann, London.

DOWSETT, R.J. (1980) The Migration of Coastal Waders from the Palaearctic across Africa, *Le Gerfant* 70, pp. 3–36.

EVANS, M.E. (1979) The Effects of Weather on the Wintering of Bewick Swans *Cygnus columbianus bewicki* at Slimbridge, England, *Scandinavian Ornithology* 10, pp. 124–132.

FARNER, D.S. (1977) Measurement of Daylength by Photoperiodic Birds, *Proceedings 1st International Symposium on Avian Endocrinology*, Calcutta University Press.

GRIFFIN, D.R. (1974) *Bird Migration*, Dover, New York.

HARRISON, C.J.O. (1979) Additional Birds from the Lower Pleistocene of Olduvai, Tanzania and Potential Evidence of Pleistocene Bird Migration *Ibis* 122, pp. 530–532.

KEETON, W.D. (1979) Avian Orientation and Navigation: a Brief Overview, British Birds 72, pp. 451–470.

LORENZ, K.Z. (1952) *King Solomon's Ring*, Methuen, London.

MATTHEWS, G.V.T. (1968 2nd Ed.) *Bird Navigation*, Cambridge University Press.

MOREAU, R.E. (1969) Comparative Weights of some trans-Saharan Migrants at Intermediate Points, *Ibis* 111, pp. 621–624.

MOREAU, R.E. (1972) *The Palearctic-African Bird Migration Systems*, Academic Press, London.

PENNYCUICK, C.J. (1969) The Mechanics of Bird Migration, *Ibis* 111, pp. 525–556.

RABØL, J. (1979) Magnetic Orientation in Night-Migrating Passerines, *Scandinavian Ornithology* 10, pp. 69–75.

RICARD, M. (1969) *Mysteries of Animal Migration*, Constable, London.

SAUER, E.G.F. (Aug. 1958) Celestial Navigation by Birds, *Scientific American.*

SCHMIDT-KOENIG, K. (1979) *Avian Orientation and Navigation*, Academic Press, London.

SEEL, D.C. (1977) Migration of the Northwestern European Population of the Cuckoo, *Cuculus canorus*, as Shown by Ringing, *Ibis* 119, pp. 309–322.

ULFSTRAND, S. (1963) Ecological Aspects of Irruptive Bird Migration in N.W. Europe, *Proceedings of the 13th International Ornithological Congress*, pp. 780–794.

## Chapter 10

ATLMANN, J. (1974) Observational Study of Behaviour: Sampling Methods, *Behaviour* 49, pp. 227–267.

ARMSTRONG, E.A. (1965) *Bird Display and Behaviour—an Introduction to the Study of Bird Psychology*, Dover, New York.

BURTT, H.E. (1967) *The Psychology of Birds*, Macmillan, London.

BURTT, H.E. (1979) Overwing and Underwing head-scratching by a male black and white warbler *Mniotilta varia*, *Ibis* 122, p. 541.

EVANS, S.M. (1970) *The Behaviour of Birds, Mammals and Fish*, Heinemann Eductional, London.

FREETHY, RON (1980) Division of Labour between Dippers Building Nests, *British Birds* 73, p. 352.

FREETHY, R. (1980) Moorhens Rapid Construction of Brood Nest, *British Birds* 73, p. 35.

FRITH, H.J. (1977) Some Display Postures of Australian Pigeons, *Ibis* 119, pp. 167–182.

GOODWIN, D. (1961) *Instructions to Young Ornithologists, Vol. 2, Bird Behaviour*, Museum Press, London.

GOODWIN, D. (1965) *Instructions to Young Ornithologists, Vol. 6, Domestic Birds*, Museum Press, London.

GOTTFRIED, B.M. (1979) Anti-predator Aggression in Birds Nesting in Old Field Habitats: an Experimental Analysis, *Condor* 81, pp. 251–257.

LORENZ, K. (1952) *King Solomon's Ring*, Methuen, London.

MANNING, A. (1972) *An Introduction to Animal Behaviour*, Edward Arnold, London.

MENDENHAL, V.M. (1979) Brooding of Young Ducklings by Female Eiders *Somateria mollissima*, *Scandinavian Ornithology* 10, pp. 94–99.

SIMON, H. (1977) *The Courtship of Birds*, Cassell, London.

STILES, F.G. AND WOLF, L.L. (1979) Ecology and Evolution of lek Mating Behaviour in the Long-tailed Hermit Hummingbird, *Ornithological Monographs* 27, pp. 1–77.

THORPE, W.H. AND ZANGWILL, O.L. (1961) *Current problems in Animal Behaviour*, Cambridge University Press.

TINBERGEN. N. (1953) *A Herring Gull's World*, Collins, London.

TINBERGEN, N. (1965) *Social Behaviour in Animals*, Chapman and Hall, London.

YASUKAWA, (1979) Territory Establishment in Red Winged Blackbirds: Importance of Aggressive Behaviour and Experience, *Condor* 81, pp. 258–264.

## Chapter 11

BANKO, E.W. (1978) *Hawaiian Bird Bibliography*, 3 parts, University of Hawaii.

BANKO, E.W. (1980) *History of Endemic Hawaiian Birds*, University of Hawaii.

BANNERMAN, D.A. (1953) *The Birds of West Equitorial Africa*, 2 vols, Oliver and Boyd, Harlow.

BLAKE, E.R. (1977) *Manual of Neotropical Birds*, Vol. 1, Chicago University Press.

BOND, J. (1971) *Birds of the West Indies*, Collins, London.

BROWN, L., URBAN, E., AND NEWMAN, K. (1981) *Birds of Africa*, Academic Press, London.

BURTON, JOHN (1981) Half North Half South: Britain's Rich Wildlife, *New Scientist* 90/1247, pp. 16–18.

CRAMP, S., SIMMONS *et al* (eds) *Birds of the Western Palaearctic*, 2 vols so far published, Oxford University Press.

CROSSKEY, R.W. AND WHITE G.B. (1977) The Afro-tropical Region: a Recommended Term in Zoo-geography, Journal of Natural History 11, pp. 541–544.

CYRUS, D. AND ROBSON, N. (1980) *Bird Atlas of Natal*, University of Natal Press.

ETCHEOPAR, R.D. AND HUL, F. (1967) *The Birds of North Africa*, Oliver and Boyd, Harlow.

FELLA, R.A. SIBSON, R.B. AND TURBOTT, E.G. (1966) *A Field Guide to the Birds of New Zealand*, Collins, London.

FRITH, H.J. (1967) *Waterfowl in Australia*, Angus and Roberton, Sydney.

GADOW, M. (1913) *The Wanderings of Animals*, Cambridge University Press.

GEORGE, W. (1962) *Animal Geography*, Heinemann, London.

GODFREY, W.E. (1966) *The Birds of Canada*, National Museum, Ottowa.

HAILA, Y., JÄRVINEN, O. AND VÄISÄNCA, R.A. (1978) Effect of Mainland Population Changes on the Terrestrial Bird Fauna of a Northern Island, *Scandinavian Ornithology* 10, pp. 48–55.

JENSON, J.V. AND KIRKEBY, J. (1980) *The Birds of the Gambia*, ARHW, Denmark.

JOHNSGARD, P.A. (1980) *Birds of the Great Plains, N. America*, Nebraska.

JOHNSON, A.W. (1965) *Birds of Chile and Adjacent Regions*, Buenos Aires University Press.

KARMALI, J. (1980) *Birds of Africa*, Collins, London.

KEAST, A. AND MORTON, E.S. (eds) (1980) *Migrant Birds in the Neotropics*, Smithsonian Institute, Washington, D.C.

LACK, D. (1976) *Island Biology Illustrated by the Land Birds of Jamaica*, Blackwell, Oxford.

LAIRD, M. (1980) *Bibliography of the Natural History of Newfoundland and Labrador*, Academic Press, London.

MACARTHUR, R.H. (1972) *Geographical Ecology: Patterns in the Distribution of Species*, Harper and Row, London.

MACKWORTH-PRAED, C.W. AND GRANT, C.H.B. (1952) *African Handbook of Birds*, Longmans, London.

MCDONALD, J.D. (1973) *Handbook of Australian Birds*, Reed, Sydney.

MOREAU, R.E. (1972) *The Palearctic-African Bird Migration Systems*, Academic Press, London.

MURPHY, R.C. (1936) *Oceanic Birds of South America*, 2 vols. American Museum of Natural History, Washington.

PALMER, R.S. (ed) (1976) *Handbook of North American Birds*, Yale University Press.

PENNY, M. (1974) *The Birds of the Seychelles and the Outlying Islands*, Collins, London.

PIZZEY, G. (1974) *Field Guide to the Birds of Australia*, Collins, London.

RIDGELY, R.S. (1976) *A Guide to the Birds of Panama*, Princeton University Press.

ROBERTS, J. (1978 edition) *Birds of South Africa*, Voeleker, Cape Town.

SCHAUENSEE, R.M. de (1970) *A Guide to the Birds of South America*, Livingstone.

SCHAUENSEE, R.M. de AND PHELPS, W.H. JR (1977) *A Guide to the Birds of Venezuela*, Princeton University Press.

SHARROCK, J.T.R. (ed.) (1976) *The Atlas of Breeding Birds of Britain and Ireland*, Poyser and BTO, Staffordshire and Tring.

SNOW, D.W. (1978) Relationships between the European and African Avifaunas, *Bird Study* 25, pp. 134–148.

SNYDER, L.L. (1957) *Arctic Birds of Canada*, University of Toronto Press.
THOMAS, D.G. (1979) *Tasmanian Bird Atlas*, University of Tasmania Press.
TOMIATOJE, L. (1976) *Birds of Poland—A List of Species and their Distribution*, Washington University Press.
VAURIE, C. (1965) *The Birds of the Palaearctic Fauna*, Witherby, London.
VOCUS, K.H. (1959) *Atlas of European Birds*, Nelson, Sunbury-on-Thames.
WHITTELL, H.M. (1954) *The Literature of Australian Birds*, Perth.
WILLIAMS, J.G. (1963) *A Field Guide to the Birds of East and Central Africa*, Collins.
WILLIAMS, J.G. (1967) *A Field Guide to the National Parks of East Africa*, Collins, London.
WILSON, EDWARD (1975) *Birds of the Antarctic*, Blandford Press, Poole.
YEATMAN, L. (ed.) (1976) *Atlas des Oiseaux Nicheurs de France*, Société Ornithologique de France, Paris.

## Chapter 12

BARBER, ED. (1970) *Farming and Wildlife—A Study in Compromise*, RSPB, Sandy.
COOPER, J.E. AND ELEY, J.T. (1979) *First Aid and Care of Wild Birds*, David & Charles, Newton Abbot.
CRAMP, S. (1977) *Bird Conservation in Europe*, HMSO, London.
DORWARD, D. (1979) *A Case of Comeback: the Cape Barren Goose*, *Australian Natural History* 19/4, pp. 130–135.
DUFFEY, E. (1974) *Nature Reserves and Wildlife*, Heinemann, London.
FREETHY, RON (1981) *The Making of the British Countryside*, David and Charles, Newton Abbot.
GOODWIN, D. (1978) *Birds of Man's World*, Cornell University Press.
HARRISON, J. AND GRANT, P. (1976) *The Thames Transformed*, Andre Deutsch, London.
HERMAN, O. AND OWEN, J.A. (1909) *Birds Useful and Birds Harmful*, Manchester University Press.
HORTON, N. (1979) *Bio-acoustic Bird Scarers (BABS)*, Civil Aviation Authority Document 157, p. 8.
KEAR, J. AND BERGER, A.J. (1980) *The Hawaiian Goose*, Poyser, Berkhamstead.
KUSHLAN, J.A. AND WHITE, D.A. (1977) Nesting Wading Bird Populations in *Southern Florida*, Florida Science 40, pp. 65–72.
MELLANBY, K. (1967) *Pesticides and Pollution*, Collins, London.
MURTON, R.K. AND WRIGHT, E.N. (1968) *The Problem of Birds as Pests*, Academic Press, London.
OLNEY, P.J.J. (Ed.) (1965) *List of European and North American Wetlands of International Importance*. Project Marc. IUCN Pub. New Series 5.
WATSON, J. (Ed.) (1893) *Ornithology in Relation to Agriculture and Horticulture*, London.
WHEELER, W. (1980) *The Thames Transformed*, Routledge and Kegan Paul, Henley.
WILLIAMS, M. (1979) The Status and Management of Black Swans, *Cygnus atratus*, at Lake Ellesmere since the 'Wahine Storm' April 1968, *New Zealand Journal of Ecology* 2, pp. 34–41.
WRIGHT, E.N., INGLIS, I.R. AND FEARE, C.J. (Eds) (1980) *Bird Problems in Agriculture*, British Crop Protection Council, London.

## Further Reading

ALLABY, M (ed) (1986) *Oxford Dictionary of Natural History*, Oxford University Press.

BOAG, D. AND ALEXANDER, M. (1986) *The Atlantic Puffin*, Blandford Press, Poole.

COLLAR, N.J. AND ANDREW, P. (1988) *Birds to Watch: The ICBP World Checklist of Threatened Birds*, Cambridge ICBP.

CROXALL, J.P., EVANS, P.G.H. AND SCHREIBER, R.W. (eds) (1984) *Status and Conservation of the World's Seabirds*, Norwich ICBP.

CRAMP, S. (ed) (1977–1987) *The Birds of the Western Palaearctic* (5 vols), Oxford University Press.

FLEGG, J.J.M. (1985) *The Puffin*, Shire Publications, Aylesbury.

FREETHY, RON (1985) *British Birds in their Habitats*, The Crowood Press, Ramsbury.

FREETHY, RON (1987) *Auks, An Ornithologist's Guide*, Blandford Press, Poole.

FREETHY, RON (1989) *Wildlife in Towns*, The Crowood Press, Ramsbury.

FULLER, E.R. (1987) *Extinct Birds*, Viking, London.

LEE, B. (1989) *Fields, Farms and Hedgerows*, The Crowood Press, Ramsbury.

LOCKWOOD, W.B. (1984) *The Oxford Book of British Bird Names*, Oxford University Press.

LOVEGROVE, R. AND SNOW, P. (1984) *River Birds*, Columbus Books, London.

MOUNTFORD, G. AND ARLOTT, B. (1988) *Rare Birds of the World*, Collins/ICBP, London.

PRATT, H.D., BRUNNER, P.L. AND BERRETT, B. (1987) *The Birds of Hawaii and the Tropical Pacific New Jersey*, Princeton University Press.

SCHAUENSEE, R.M. DE (1984) *A Guide to the Birds of South America*, Academy of Natural Sciences, Philadelphia (reprinted by Pan American section ICBP).

SCHAUENSEE, R.M. DE (1984) *The Birds of China*, Oxford University Press.

STUART, S.N. AND JOHNSON, T. (eds) *World Checklist of Threatened Species*, Nature Conservancy Council, London.

WHITE, C.M.N. AND BRUCE, M.D. (1986) *The Birds of Wallacea*, British Ornithologists Union.

# Illustration Credits

Figs 1a, 1b, British Museum (Natural History); Figs 2, 3, 4, 5, 6, 7, Ray Hutchins; Figs 8, 9, Ron Freethy; Figs 10, 11, Ray Hutchins; Figs 12, 13, *Encyclopaedia of Aviculture*, Blandford Press; Fig 14, Ray Hutchins; Figs 15, 16, *Encyclopaedia of Aviculture*, Blandford Press; Fig 17, Doreen Edmond; Figs 18, 19, *Encyclopaedia of Aviculture*, Blandford Press; Fig 20, Ray Hutchins; Fig 21, *Encyclopaedia of Aviculture*, Blandford Press; Fig 22, Ron Freethy; Fig 23, Julian Fitter/Oxford Scientific Films; Fig 24, Ron Freethy; Figs 25, 26, Ray Hutchins; Fig 27, Eric Soothill; Fig 28, Ron Freethy; Fig 29, Richard Millington; Figs 30, 31, Ron Freethy; Figs 32, 33, *Encyclopaedia of Aviculture*, Blandford Press; Fig 34, Eric Soothill; Fig 35, Mike Read and Martin King; Fig 36, Ron Freethy; Figs 37, 38, 39, 40, 41, *Encyclopaedia of Aviculture*, Blandford Press; Figs 42, 43, Richard Millington; Fig 44, *Encyclopaedia of Aviculture*, Blandford Press; Fig 45, J.A.L. Cooke/Oxford Scientific Films; Fig 46, Doreen Edmond; Fig 47, *Encyclopaedia of Aviculture*, Blandford Press; Figs 48, 49, Richard Millington; Figs 50, 51, 52, 53, 54, Ray Hutchins; Fig 55, London Scientific Fotos; Figs 56, 57, 58, 59, 60, 61, 62, 63, 64, 65, 66, 67, 68, 69, 70, 71, Ray Hutchins; Fig 72, Ron Freethy; Figs 73, 74, 75, Ray Hutchins; Fig 76, Ron Freethy; Figs 77, 78, Ray Hutchins; Fig 79, Ron Freethy; Fig 80, Ray Hutchins; Fig 81, Ron Freethy; Figs 82, 83, Ray Hutchins; Fig 84, Mike Mockler; Figs 85, 86, 87, Ray Hutchins; Fig 88, *The Meaning of Bird Sounds*, Jellis, R., Collins; Fig 89, G.L. Kooyman/Oxford Scientific Films/Animals Animals; Figs 90, 91, Ray Hutchins; Figs 92, 93, Ron Freethy; Fig 94, Mike Mockler; Fig 95, Eric Soothill; Fig 96, Mike Mockler; Figs 97, 98, 99, 100, Ray Hutchins; Fig 101, Ron Freethy; Fig 102, Ray Hutchins; Fig 103, Ron Freethy; Fig 104, Richard Littleton; Fig 105, Ray Hutchins; Fig 106, Stouffer Enterprises Inc./Oxford Scientific Films/Animals Animals; Fig 107, Jane Burton; Fig 108, Ray Hutchins; Fig 109, Eric Soothill; Fig 110, Ron Freethy; Fig 111 *left* Raymond Parlett/Halcyon Photographic Library; Fig 111 *right*, Christopher Ware/Halcyon Photographic Library; Fig 112, Pat Lee; Fig 113, Michael W. Richards/RSPB; Fig 114, Mike Read and Martin King; Fig 115, Ron Freethy; Fig 116, British Trust for Ornithology; Fig 117, Ron Freethy; Endpapers, Esao Hashimoto/Oxford Scientific Films/Animals Animals.

Colour Plates: I, Ron Freethy; II, Godfrey Merlen/Oxford Scientific Films/Animals Animals; III, Leonard Lee Rue III/Oxford Scientific Films/Animals Animals; IV, Brian Milne/Oxford Scientific Films/Animals Animals; V, Alan G. Nelson/Oxford Scientific Films/Animals Animals; VI, Mike Read and Martin King; VII, C.M. Perrins/Oxford Scientific Films/Animals Animals; VIII, G.I. Bernard/Oxford Scientific Films; IX, Alan G. Nelson/Oxford Scientific Films/Animals Animals; X, Charles Palek/Oxford Scientific Films/Animals Animals; XI, J.K. Burras/Oxford Scientific Films; XII, G.L. Kooyman/Oxford Scientific Films/Animals Animals.

# Index of Common Names